A NATURALIS

INSECTS OF AUSTRALIA

Peter Rowland and Rachel Whitlock

JOHN BEAUFOY PUBLISHING

This edition published in the United Kingdom and Australia in 2022 by John Beaufoy Publishing Ltd
11 Blenheim Court, 316 Woodstock Road, Oxford OX2 7NS, England
www.johnbeaufoy.com

Photo Credits
Front cover: *main image:* Teddy Bear Bee © Jenny Thynne; bottom row, *left to right:* Common Ellipsidion © Andrew Allen, Cruiser (male) © Peter Rowland/kapeimages.com.au, Potter Wasp © Andrew Allen.
Back cover: Giant King Cricket © Jordan de Jong.
Title page: Golden Green Stag Beetle © John Nielsen.
Contents page: Queensland Fruit Fly © Peter Rowland/kapeimages.com.au

ISBN 978-1-913679-26-2

Edited by Krystyna Mayer
Designed by Gulmohur Press, New Delhi
Printed and bound in Malaysia by Times Offset (M) Sdn. Bhd.

·Contents·

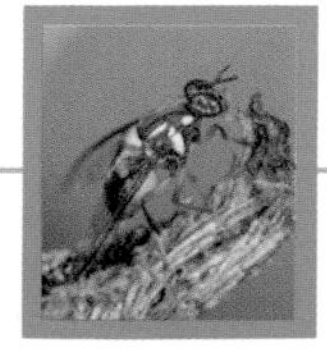

Introduction

Insects belong to the largest phylum of animals, the arthropods (Arthropoda), which are invertebrates (lack a backbone) and are characterized by having an exoskeleton, segmented body and paired jointed appendages. They belong to the class Insecta, by far the largest class of any animal, with more than a million documented species worldwide. Adult insects are the only invertebrates to possess wings, although these are not always present, and are further characterized by having three pairs of legs, a pair of antennae and three generally distinct body segments: head, thorax and abdomen. Primitive insects were wingless, with ancestors presumably much like the bristletails and silverfish of today. The development of wings some 400 million years ago gave insects the power of flight. Early winged insects had fixed, outstretched wings much like those of today's dragonflies, and over the next 65 million years came the ability to fold back the wings over the abdomen.

Given the huge number of insect species found in Australia, it is beyond the scope of this book to include them all. Instead, the aim is to include the insect genera and some representative species that are most commonly found or most broadly distributed in the eastern parts of Australia, from around Darwin in the Northern Territory, through Queensland, New South Wales, Australian Capital Territory, Victoria and Tasmania, to Adelaide in South Australia. This is not, therefore, a comprehensive field guide, but an introduction to the key information required by an amateur naturalist when viewing the species included in their natural environments.

Modern Australian Insects

About 1,000 modern insect families are widespread throughout the world; of these more than 700 are found in Australia, including over 13,200 genera and over 60,000 described species. For classification to family level, including the total number of genera and species, see the checklist (p. 158). An introduction to the 25 Australian insect orders is included below. This follows current classification in the Australian Faunal Database (http://biodiversity.org.au/afd).

Bristletails (Order: Archaeognatha) Bristletails have a wingless, cylindrical body with fine scales, large compound eyes and three elongated, rear-facing tail appendages, or cerci, the central cercus longer than the outer ones. They are superficially similar to silverfish, which have outwards-facing outer cerci and smaller eyes. Bristletails can jump long distances using their cerci, unlike silverfish.

Silverfish (Order: Zygentoma) Silverfish have a wingless, flattened body with or without scales, small compound eyes (sometimes absent), and three cerci, the central one longer than the outer pair. The central cercus is rear facing, while the outer ones face sideways, a characteristic that distinguishes them from the bristletails.

Dragonflies and Damselflies (Order: Odonata) These are subdivided into two suborders: the dragonflies of Odonata, and the damselflies of Zygoptera. The body is long and thin with a large head and large compound eyes. There are two pairs of long membranous wings, the hindwings broader than the forewings, and held horizontal in dragonflies and typically rear facing in damselflies when at rest. Both the aquatic nymphs and flying adults are active predators.

Stick and Leaf Insects (Order: Phasmida) The body is typically very long and cylindrical, resembling sticks, or flattened, resembling leaves, with long, thin or flattened legs. Some Australian species can exceed 300mm in body length and are among the longest insects known. The wings, when present, consist of larger membranous hindwings protected partially under shorter, hardened forewings, or tegmina. Most species rely on their excellent camouflage for protection and are easily overlooked in their natural habitats, rarely fleeing from predators; instead they remain motionless in situ or after dropping to the ground, or spread their large hindwings in an attempt to startle would-be attackers.

Caddisflies (Order: Trichoptera) The body is elongated and there are two pairs of variably sized, hairy wings, the forewings larger than the hindwings in some species, but the other way round in others. The antennae are long and filamentous, curled in some species, and the mouthparts are simple, lacking the curled proboscis found in the superficially similar moths. Larvae are aquatic, with developed legs and mandibles, and typically live in a case made of silk with various materials attached to the outside.

Praying Mantises or Mantids (Order: Mantodea) Species vary greatly in size, at 10–120mm, but all are characterized by forelimbs held forwards in 'praying' pose, each having tibia adorned with 1–2 rows of spines. All species are predatory, using their large eyes and mobile head to locate prey before seizing it with the forelimbs and using the mandibulate mouthparts to consume it. The hindwings are membranous, concealed when not in use under hardened forewings. Most species hunt on vegetation. Eggs are laid in masses, protected within a hardened foam case, or ootheca.

Mayflies (Order: Ephemeroptera) Most mayflies have two pairs of membranous wings, and forewings larger than hindwings (when present), but still only of moderate length. Adults have non-functioning mouthparts (mandibulate in nymphs), large eyes and three ocelli, moderately short antennae and three (rarely two) long, filamentous cerci. Nymphs are aquatic and look much like wingless adults.

Stoneflies (Order: Plecoptera) The body is flattened, with two pairs of membranous wings, the forewings slightly longer and narrower than the hindwings, and held curved around the body when at rest. The legs are long and face forwards when not moving. The elongated abdomen has two cerci. Larvae are aquatic.

Cockroaches and Termites (Order: Blattodea) Several introduced species are common pests around residential and commercial buildings. These feed largely on human foodstuffs and waste and are known to spread a number of dangerous diseases to humans. The native species are not considered dangerous to humans, living in the ground litter and under tree bark. Cockroaches have a generally broad body and pronotum, or frontal shield, partly covering the head, which has downwards-facing mandibulate mouthparts. All nymphs and some adults lack wings, but the wings of species with winged adults are membranous, the forewings toughened and overlapping when not in use. Termites form large multicaste colonies, with a single reproductive pair (queen and reproductive male), workers, soldiers and nymphs, in timber or in constructed small to large mounds. They are subterranean or arboreal, feeding on dead grass and timber. Most members of a colony are wingless, with a winged reproductive caste, or alates, appearing only to disperse away from the parent colony to breed and establish new colonies.

Webspinners (Order: Embioptera) These small insects are uncommonly encountered. Males usually have two pairs of membranous wings, the forewings larger than the hindwings, while females are wingless. Males also have short cerci of differing lengths. The body is elongated and tube-like, with three pairs of legs, the front pair facing forwards and the other two pairs facing rearwards. The characteristic boxing glove-like bulbous tarsi on the front legs are used to produce silk. Females form small colonies with other females and combined offspring of recent matings, concealed in silk tunnels.

Grasshoppers, Crickets and Katydids (Order: Orthoptera) This large order contains two suborders: Caelifera, containing grasshoppers, and Ensifera, containing crickets and katydids. Species can attain large sizes (to 100mm), and most possess big hindlegs that give them a powerful jumping ability, with some using the legs to produce their song. Wings, not present in all species, typically consist of membranous hindwings that are folded under protective forewings, although some winged species have soft forewings. The head usually has large eyes and mandibulate mouthparts. Ensifera females possess an ovipositor, while those of the Caelifera do not. The Caelifera also have shorter antennae with fewer segments (<30).

Earwigs (Order: Dermaptera) The tip of the abdomen has pronounced forcipules/pincers, which are curved in males and straight with inwards-pointing tips in females. Wings, when present, consist of reduced protective forewings and membranous hindwings folded neatly underneath. The body is generally shiny, brown to blackish and flattened. Males display to females before mating, and eggs are laid in batches in soil crevices and similar sites, which are guarded against predators and cleaned regularly.

Zorapterans (Order: Zoraptera) Zorapterans are small (to 3mm), elongate insects that superficially resemble termites. Winged morphs have small eyes and shed their wings after dispersing from a colony, leaving small stubs, while wingless, or apterous, morphs are blind and live in a colony. The mouthparts are mandibulate, the antennae segmented and of moderate length, and the cerci short and often unequal in size and shape in males, with a distinctive hook. The colour is typically off-white to brown. Zorapterans feed mainly on fungal hyphae and spores in rotting timber and leaf litter, but have also been recorded feeding on nematodes, mites and other minute arthropods.

Booklice, Barklice, Biting and Sucking Lice (Order: Psocodea) Booklice and other psocid lice can be wingless or winged (membranous), with a large head, biting mouthparts and protruding eyes at the rear of the sides of the head. While a small number may feed on books and similar stored cellulose products, the vast majority ingest minute organic particles on vegetation under fallen timber or rocks, or in the soil. Biting and sucking lice are specialized external parasites of birds and mammals, relying on their hosts for their survival. These lice are wingless, with a dorsoventrally flattened body, biting or sucking mouthparts, and very short, stubby antennae. The eyes are either greatly reduced in size or absent.

Thrips (Order: Thysanoptera) Thrips are very small (0.5–15mm), with an elongated, cylindrical body and three pairs of legs, each of which ends in a specialized bladder-like organ instead of tarsal claws. Wings, when present, are membranous and hairy, the forewings larger than the hindwings. Parthenogenesis is common, with females produced from unfertilized eggs. Most thrips feed on pollen and flowers, and a few on fungi. A small number of mainly introduced species can be pests of commercial orchards and gardens.

True Bugs (Order: Hemiptera) True bugs are very variable in both size (1–110mm) and colour, with some species being among the most brightly coloured and spectacularly patterned of all insects. Wings, if present, cross over when at rest, and wingless species typically live under protective coverings. All species have sucking, tube-like mouthparts. Some families can be direct pests of humans; bed bugs are external parasites of humans, and assassin bugs can carry disease-causing pathogens. Others, such as aphids, are pests of crops and flower gardens.

Lacewings, Antlions and Mantidflies (Order: Neuroptera) Adults have long, heavily veined, membranous wings and weak, fluttering flight. The mouthparts are mandibulate. Larvae are wingless, and are predators of other invertebrates. Some create a pit trap in loose, dry soil, concealing themselves under the base of the trap, quickly grabbing prey animals that stray within striking distance with powerful, forcep-like, hollow jaws, and dragging the victim underground, where it is sucked dry. Most actively roam in search of prey, some camouflaging themselves with the body parts of their prey.

Alderflies, Dobsonflies and Fishflies (Order: Megaloptera) All have two pairs of large (to 100mm), similarly sized, membranous wings, mandibulate mouthparts and slender legs. Wings are typically patterned. Larvae are aquatic and predate on other aquatic invertebrates. Adults live in the vicinity of water.

Beetles (Order: Coleoptera) Beetles are very variable in size (0.4–80mm) and colour, but are usually typified by modified, hardened forewings, or elytra, designed to cover and protect the membranous hindwings, which are folded underneath when at rest. The elytra form a straight midline down the body when at rest and are not used in flight, but held away from the beating hindwings. In some species the forewings are reduced and some species are wingless. All stages of development have mandibulate mouthparts. Sexual dimorphism is apparent in many species, and extreme variations in the shapes of body parts can occur.

Twisted Wings and Stylops (Order: Strepsiptera) Males possess enlarged, membranous hindwings with little to no visible venation, and forewings reduced to long appendages. They have three pairs of legs and branched antennae. Females are wingless and often have no legs. Females are endoparasites of other insects, particularly beetles and wasps, clinging to their host with small mouthparts and typically remaining with it for their whole life; the mouthparts of males are reduced and non-functional.

Flies, Mosquitoes and Gnats (Order: Diptera) All have a single pair of membranous forewings. The hindwings are reduced to small, 'club-like', stabilizing appendages called halteres. All have compound eyes and most have sucking mouthparts of various designs, but the mouthparts can also be piercing (as in mosquitoes). Adult females of most mosquito species are well known for sucking the blood of animals (including humans) to obtain the proteins required for egg production, and in doing so have become a major vector of diseases. Males typically feed on nectar. Other fly families also require blood for this purpose, including the tabanid flies and biting midges.

Scorpionflies and Hanging Flies (Order: Mecoptera) Both sexes have strongly veined, membranous wings, both pairs of about equal size, with the exception of the genus *Apteropanorpa*, which are wingless. The legs are long and slender, equipped with claws, and the head has an elongated projection with mandibulate mouthparts at the tip. Males of some species have an upturned, 'scorpion-like' terminal segment on the abdomen. Some species predate on other insects, using their hooked claws to seize passing prey, while others are scavengers of dead or dying animals or are herbivores.

Fleas (Order: Siphonaptera) The mouthparts are piercing and sucking, and both males and females feed on blood, sucking it from their hosts by cutting through the skin, injecting the wound with an anticoagulant and siphoning up the blood. They can survive for several months without a blood meal. They are small (to 10mm), wingless, with laterally compressed bodies, and are well known for their strong jumping ability, powered by the greatly elongated hindlegs. Almost all fleas feed on mammal blood.

Butterflies and Moths (Order: Lepidoptera) Butterflies and moths are superficially similar, but butterflies differ from moths basically by being diurnal (although some moths are also diurnal), having clubbed antennae (often hairy or feathery in moths, but can be clubbed in some species, such as *Synemon* spp.), and by generally holding their wings upright when perched (often outspread or domed over the back in moths). Both groups have compound eyes and a coiled proboscis. Larval caterpillars have three pairs of legs on the thorax and hook-tipped prolegs on most other body segments.

Bees, Wasps, Ants and Sawflies (Order: Hymenoptera) Most species in this order belong to the 'true' bees, wasps and ants of the suborder Apocrita. The abdomen is divided into two sections with the second segment reduced to form a 'waist', or petiole. There are two pairs of membranous wings, the forewings longer than the hindwings, large compound eyes and mandibulate mouthparts. Female sawflies (suborder Symphyta) use the saw-like ovipositor to cut into the leaf or stem of a host plant to lay eggs; they lack an obvious narrow 'waist' between the thorax and abdomen. Sawflies also have more complex wings than other members of this order.

NON-INSECT HEXAPODS

This group of hexapods was once classified as three orders in the class Insecta. Current classification comprises two classes: the springtails of the class Collembola, and the proturans and diplurans (two-pronged bristletails) of the class Entognatha. With the Insecta, the Collembola and Entognatha form the invertebrate subphylum Hexapoda. All have six legs and mouthparts that are somewhat internal, due to being obscured by parts of the face (entognathous). The life cycle consists of an egg and several instars between this and the adult stage.

Springtails (Class: Collembola) Springtails are generally tiny (body typically 0.5–5mm long), but can rarely grow to about 10mm. They are wingless, with a cylindrical or almost spherical body, and eyes comprising groups of ocelli. Their common name comes from the jumping mechanism located at the tip of the abdomen, either fork shaped or a singular prong. They are commonly seen around decaying vegetable or animal matter, where they feed on microorganisms.

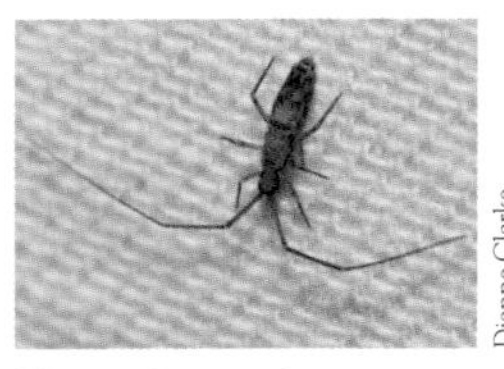

Dianne Clarke

Elongate Springtail (Entomobryomorpha)

Proturans (Order: Protura) The proturans have an elongated, cylindrical, tiny body, less than 2mm long, and lacking colouration. There are three pairs of legs, the front pair often projecting forwards, but wings, eyes, antennae and cerci are absent. They are soil dwelling, mainly in forests and woodland, where they probably feed on fungi. They are not easily sighted and species are very difficult to tell apart.

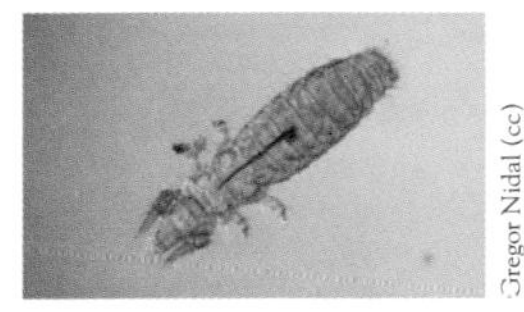

Gregor Nidal (cc)

Protura

Two-pronged Bristletails (Order: Diplura) The abdomen is tipped with two short cerci, which are pincer-like in a few species, and long and filamentous in others. The body is elongated and cylindrical, and can vary greatly in length between species, at 4–50mm. The antennae are usually long, with multiple segments, and eyes are absent. The Diplura are soil dwelling, mostly feeding on plant material, but some species are predatory, hunting and seizing prey with the pincer-like cerci.

Steve and Alison Pearson

Campodea sp. (Diplura)

LIFE CYCLES

Depending on the order to which an insect belongs, it goes through one of two types of life cycle (metamorphosis) on its journey from egg (not present in all species) to adult. **Incomplete metamorphosis** is a life cycle that has several nymphal stages after the egg hatches, which are small, wingless versions of the adult, with the adult form attained after the final nymphal moult. **Complete metamorphosis** comprises four stages: egg (may be absent), larva, pupa and adult. The larvae are markedly different from the adult form, often occupying completely different habitats and having different diets. Insects that go through complete metamorphosis include ants, bees, butterflies, flies, moths and wasps, while those that go through incomplete metamorphosis include cockroaches, mantises, stick insects, termites and true bugs. The wingless insects (Apterygotes) go through ametabolous development, with no significant metamorphosis.

ABBREVIATIONS

DISTRIBUTION KEY
ACT Australian Capital Territory; **NSW** New South Wales; **NT** Northern Territory; **NZ** New Zealand; **PNG** Papua New Guinea; **Qld** Queensland; **SA** South Australia; **Tas** Tasmania; **Vic** Victoria; **WA** Western Australia.

KEY FEATURES AND MEASUREMENTS
Sizes quoted in species accounts for body measurements (where available) are average maximum sizes, but exceptions can occur. **BL** body length; **TL** total length; **WS** wingspan.

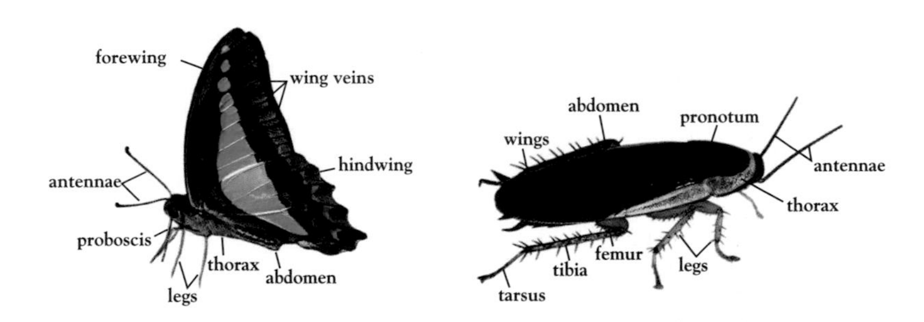

General Insect Anatomy

GLOSSARY

anaphylaxis Severe fast-acting allergic reaction that can be fatal.
apex/apical area The tip of a butterfly's forewing, where the costa meets the termen.
aposematic Refers to conspicuous pattern or bright colouration of an organism, warning predators of its toxicity or unpalatable taste.
caudal At or near back half of body.
cerci Paired appendage on abdomen, used for sensory functioning.
disc/discal area The cell or central band of a butterfly's wing that is relatively free of veins.
elytra Modified hardened forewing.
endemic Found only in a certain area.
eusocial Highest level of social organization in animals, with overlapping generations fulfilling different roles and responsibilities.
exoskeleton External skeleton used as protection.
family Taxonomic rank above genus and below order.
genus (pl **genera**) Taxonomic group above species and below family.
gregarious Living within a group or community.
haltere(s) Modified hindwing, forming small, 'club-like' balancing organ.
instar Developmental stage in insect larva between moulting periods.
invertebrate Animal that lacks a backbone.
larva Newly hatched, wingless stage of an animal.
lunule(s) Crescent-shaped part or marking on a butterfly's wing.
mandibles Jaw-like mouthparts used for grasping and cutting.
median Central part of a butterfly's wing, midway between the base and the apex.
mesonotum Upper surface of second thoracic segment.
nymph Immature insect, newly hatched to subadult stages.
ootheca Egg case of preying mantises, cockroaches and other insects.
order Taxonomic rank above family and below class.
ovipositor Specialized organ on abdomen of some female insects; used for depositing eggs.
palp Sensory organ; part of insect mouthparts.
pheromone Chemical released to communicate with members of the same species.
postdiscal The area, or band, of a butterfly's wing between the discal and the marginal area.
postmedian The central area of a butterfly's wing beyond the median and towards the apex.
pterostigma Small, non-transparent section near tip of leading edge of some insects' wings.
pupate To develop into a pupa.
sex brands Markings on the upperside of a male butterfly's forewings indicating the location of pheromone-releasing cells (androconia).
species Basic unit of taxonomic classification.
subapical Located below the apex of a butterfly's wing.
submedian Cell below the median area of a butterfly's wing.
subspecies Level of taxonomic division below species.
subtornal Located below the tornus of a butterfly's wing.
symbiotic Close, mutually beneficial relationship between two different organisms.
termen Outer margin of a butterfly's wing, furthest from the body.
tornus/tornal area The inner rear corner of a butterfly's wing.
tubercle Rounded, raised nodule.

MEINERTELLIDAE (BRISTLETAILS)

Bristletails ■ *Multiple species* TL <20mm

DESCRIPTION Wingless, grey to brown, elongated body, covered in scales. Long antennae. Compound eyes meet in middle of head, and 3 appendages at tip of abdomen; middle filament much longer than 2 cerci on either side. **DISTRIBUTION** Throughout Australia. **HABITAT AND HABITS** Found in damp environments in a variety of habitats, often under bark, leaf litter or rocks, or in soil. Nocturnal, feeding on lichens, algae, decaying wood and other vegetable matter; may also scavenge on dead invertebrates. Capable of jumping up to 10cm when disturbed.

Kristi Ellingsen

LEPISMATIDAE (SILVERFISH)

Common Silverfish ■ *Lepisma saccharina* TL 12–25mm

DESCRIPTION Flattened, usually silvery-grey, scaled, wingless insect. Long antennae and 3 long filaments at tip of abdomen that are often broken or missing. **DISTRIBUTION** Worldwide. Introduced to Australia. **HABITAT AND HABITS** Commonly found in human habitation in damp environments. May occur in kitchens, feeding on food scraps or cereal. Can become a pest when eating paper products and starchy glue in book bindings and wallpaper. Can survive long periods without food or water, and is mostly nocturnal. Also found outdoors, under rocks, bark and leaf litter.

Donald Hobern (cc)

Aeshnidae (Hawkers and Emperors)

Australian Emperor ■ *Anax papuensis* TL 70mm; WS 100mm

DESCRIPTION Adults pale brown to yellow, mottled with darker brown. Larvae elongated and large (to 45mm) with smooth body. **DISTRIBUTION** Throughout Australia, including offshore islands. Also NZ, New Caledonia, New Guinea and southern Indonesia. **HABITAT AND HABITS** Occurs near still fresh water, but may also enter slow-moving streams. Adults are active hunters of other flying insects. Does not seem to form territories, and large numbers of both sexes congregate together. Male and female remain joined throughout egg-laying process, with female making slits in plants, either above or below water level, in which she lays her eggs. Larvae aquatic, feeding on small invertebrates, and can take up to 12 months to transform to adult form.

Graham Winterflood (cc)

Corduliidae (Emeralds and Green-eyed Skimmers)

Tau Emerald ■ *Hemicordulia tau* TL 50mm

DESCRIPTION Adults predominantly black with yellowish markings, lacking metallic green colour on head in other emerald dragonflies. Inverse black 'T' mark in front of eyes. Wings membranous, with yellowish-brown veins; hindwings slightly shorter and wider than forewings. Abdomen thicker in middle in males and evenly tapered in females. **DISTRIBUTION** Most of Australia, including Tas. Some historic unvalidated records from wider in Pacific. **HABITAT AND HABITS** Occurs in a variety of freshwater habitats. Occasionally inhabits rivers and streams, but more commonly associated with lagoons, swamps, pools and ponds. Adults are active fliers, seldom resting during the day but hanging vertically high in trees at night with wings outstretched. Larvae (about 24mm) aquatic, tolerating wide range of water temperatures and feeding on aquatic invertebrates.

Kristi Ellingsen

GOMPHIDAE (CLUBTAILS)

Twin-spot Hunter ■ *Austrogomphus (Austroepigomphus) praeruptus* TL 45mm; WS 60mm

DESCRIPTION Adults black with reduced yellow markings, and green, widely separated eyes. Wings membranous and transparent, each with elongated black spot (pterostigma) on front edge towards tip. Abdomen conspicuously thickened towards tip. Classification uncertain and changed a number of times in recent years; some authorities assign it to genus *Austrogomphus* with *A. melaleucae*, the latter formerly considered a separate species but now as a junior synonym. **DISTRIBUTION** Eastern and south-eastern Australia, mainly from south of Cape York Peninsula, Qld, to south-eastern NSW. Records also from central Vic and south-east SA. **HABITAT AND HABITS** Found in and around slow-moving rivers and streams, and associated pools, where adults often perch on small branches of adjacent trees. Larvae aquatic, feeding on other aquatic invertebrates.

Reiner Richter

LIBELLULIDAE (SKIMMERS AND PERCHERS)

Red Swampdragon ■ *Agrionoptera insignis* TL 40mm; WS 70mm

DESCRIPTION Adults have black-and-yellow body (synthorax), with short, thin, mostly reddish-orange abdomen with black tip. Wings membranous, each with metallic green patch towards tip and yellowish wash at base, somewhat more extensive in female. Larvae yellowish-brown, to *c.* 17mm in length. **DISTRIBUTION** Large number of widely ranging subspecies. *A. i. allogenes* in northern and eastern Australia, including NT, Qld and northern NSW. Occurs further north into New Guinea, Solomon Islands and potentially New Caledonia. **HABITAT AND HABITS** Inhabits shaded freshwater swamps, pools, ponds and streams in lowland forests, and can tolerate degrees of habitat disturbance. Larvae aquatic, feeding on other aquatic invertebrates.

Craig Nieminski

Red Percher ■ *Diplacodes bipunctata* TL 32mm; WS 55mm

Andrew Allen

Andrew Allen

TOP Male; BOTTOM Female.

DESCRIPTION Adult male orange to reddish; female yellowish-brown to greyish-brown; both have darker markings along abdomen. Wings membranous and clear, with yellowish-brown patch at base of hindwing and dark veins. Two distinct dark spots on sides of thorax distinguish it from similar species. Larvae grow to *c.* 13mm. **DISTRIBUTION** Throughout Australia except Tas, mainly below 1,000m. Also NZ and islands of Indonesia. **HABITAT AND HABITS** Occurs in and around still and slow-moving fresh water, including lakes, billabongs, swamps and temporary roadside pools, where adults readily perch on the ground and, less often, on low vegetation, making brief flights between landings. Larvae aquatic. Adults can disperse significantly away from water. Eggs laid in flight, with male and female still in embrace.

Scarlet Percher ■ *Diplacodes haematodes* TL 35mm; WS 60mm

Lorraine Harris

Male (top) and female mating.

DESCRIPTION Adult male brilliant scarlet with tapered abdomen; female yellowish-brown to greyish-brown, very similar to the Red Percher (see above), but slightly larger. Wings membranous. Tips of female's wings brownish-yellow; bases of male's hindwings yellow. Larvae grow to *c.* 14mm. **DISTRIBUTION** Throughout mainland Australia. Further north to New Guinea, Timor, New Caledonia and Vanuatu. **HABITAT AND HABITS** Inhabits still and moving water, including rivers, streams, lakes and dams. Adults readily perch on the ground and less often on low vegetation, making brief flights between landings. Males often perch on rocks near water on the lookout for females. Larvae aquatic, feeding on other aquatic invertebrates.

Blue Skimmer ■ *Orthetrum caledonicum* TL 45mm; WS 70mm

DESCRIPTION Mature adult male powdery-blue (pruinescent), with darker tip to abdomen and dark reddish-black head. Female and younger male yellow and black with greenish eyes. Wings membranous, with dark veins and metallic sheen towards tips. Larvae creamish to brown, with broad abdomen; reach *c.* 24mm in length. **DISTRIBUTION** Widely distributed throughout Australia, including Tas (less common). Further north to New Guinea, New Caledonia and Lesser Sunda Islands. **HABITAT AND HABITS** Inhabits wide range of still and flowing waters, including lakes and ponds. Males typically perch on sunlit rocks and earth at edges of a waterbody. Males seemingly territorial, readily chasing away rival males and other dragonfly species. Eggs laid in water by female during hovering flight. Larvae aquatic.

Peter Rowland/kapeimages.com.au

Slender Skimmer ■ *Orthetrum sabina* TL 50mm; WS 80mm

(Green Marsh Hawk)

DESCRIPTION Adults greenish-yellow to pale greyish with broad black markings, and green eyes. Wings membranous with small black spot at base of hindwings; wings held extended at rest. Abdomen long and slender with enlarged tip. Larvae have spines on abdominal segments and reach *c.* 21mm in length. **DISTRIBUTION** Throughout northern and eastern Australia, from around the Kimberleys, WA, through NT, Qld and NSW, to around Vic border. Also PNG, Indonesia, Asia, Middle East and North Africa. **HABITAT AND HABITS** Occurs in and around shallow, still and slow-moving waters, including temporary pools. Adults perch for long periods on vegetation and take to the air to seize other insects, including small butterflies and other dragonflies. Larvae aquatic and prey on other aquatic invertebrates, especially mosquito larvae.

John Tann (cc)

Iridescent Flutterer ■ *Rhyothemis braganza* WS 85mm

DESCRIPTION Adults blackish with broad, membranous wings, each with dark, metallic purplish-black basal patches of similar length and transparent tips. Wings lack pale spots found in similar species, and have dark pterostigma towards tip of leading edge of each wing. Eyes dark red. Larvae have mid-dorsal spines and very few setae on body. **DISTRIBUTION** Endemic to northern and north-eastern Australia, from around Port Hedland, WA, through NT and Qld, to near NSW border. **HABITAT AND HABITS** Inhabits rivers, streams and associated pools. Adults have a fluttering flight and often land on small twigs or tall grass stems. Larvae aquatic, feeding on other aquatic invertebrates.

Peter Rowland/kape.mages.com.au

Graphic Flutterer ■ *Rhyothemis graphiptera* WS 85mm

(Banded Flutterer)

DESCRIPTION Mature adults have yellowish-brown wings, each with large, dark metallic brown patches (not at tips). Abdomen greyish-brown above, paler below, and eyes red. Larvae grow to *c.* 15mm. **DISTRIBUTION** Throughout much of mainland Australia; presumed absent only from south-western WA, south-western Qld, western NSW, Vic and south-eastern SA. Also New Caledonia and Indonesia. **HABITAT AND HABITS** Occurs in and around lagoons, billabongs, ponds and swamps, where adults frequently land on tips of tall grasses. Insect prey taken in fluttering flight. Male and female remain joined throughout egg-laying process, during which female deposits eggs on the water's surface. Larvae aquatic, voraciously feeding on other aquatic insects.

Thomas Rowland/kapeimages.com.au

LINDENIIDAE (TIGERS)

Australian Tiger ■ *Ictinogomphus australis* TL 70mm; WS 100mm

DESCRIPTION Mature adults conspicuously banded and striped with blackish and pale yellow, and with enlarged tip to abdomen. Eyes large, greenish to greyish, and widely spaced. Wings membranous and clear, with elongated patch (pterostigma) towards tip of leading edge on each wing. Larvae have broad body, slightly wider than long, growing to *c*. 26mm. **DISTRIBUTION** Throughout northern and eastern Australia, from around Broome, WA, through northern NT, northern and eastern Qld, and eastern NSW, to around Vic border. **HABITAT AND HABITS** Inhabits rivers, lagoons, billabongs, lakes and ponds. Males perch conspicuously on emerging aquatic vegetation or similar exposed vantage points and drive away competing males. Female lays eggs by dipping tip of abdomen in water's surface with male in close proximity. Larvae aquatic.

John Tann (cc)

COENAGRIONIDAE (NARROW-WINGED OR POND DAMSELFLIES)

Pygmy Wisp ■ *Agriocnemis pygmaea* TL 20mm

DESCRIPTION Adult colour variable. Typically green-and-blackish head and thorax; female has reddish abdomen, while male's is metallic greenish-black and blue with reddish tip. Wings membranous, clear, with faint metallic sheen above and small black patches on front edge, towards tip of each. Wings about three-quarters length of abdomen. Larvae have numerous hair-like bristles and grow to *c*. 16mm. **DISTRIBUTION** Throughout northern and eastern Australia, from around Port Hedland, WA, through northern NT, northern and eastern Qld, and eastern NSW, to around Vic border. More widely distributed in central and southern Asia. **HABITAT AND HABITS** Found in a range of wetland habitats, both natural and artificial, including swamps and ponds. Commonly seen at rest on aquatic plants with wings held longitudinally above abdomen. Larvae aquatic, typically foraging around aquatic plants and algae.

John Tann (cc)

Common Bluetail ■ *Ischnura heterosticta* TL 34mm; WS 45mm

DESCRIPTION Adult male black with blue around head and thorax, and blue tip to olive-brown abdomen. Female more uniform olive-brown to greenish, darker above. Wings membranous, clear, with metallic bronze sheen above and small black patches towards tip of each. Wings about three-quarters length of abdomen. Larvae dark brown and grow to *c.* 17mm. **DISTRIBUTION** Widespread throughout Australia, including Tas. Further widely distributed in Indo-Pacific. **HABITAT AND HABITS** Occurs in and around still and slow-moving waterbodies. Wings held longitudinally above abdomen when at rest. Male and female remain joined throughout egg-laying process. Larvae aquatic, often seen on macrophytes, and seemingly tolerant of saline water.

Sam Fraser-Smith (cc)

Gold-fronted Riverdamsel ■ *Pseudagrion aureofrons* TL 35mm

DESCRIPTION Adults blackish and blue, with abdomen bronzed black on upper surface of segments 2–4 and blue at tip. Adult male has yellowish face and thorax. Wings membranous, clear, with metallic bronze sheen above and small black patches towards tip of each. Larvae greyish-black and elongate, growing to *c.* 22mm. **DISTRIBUTION** Throughout most of Australian mainland from around Carnarvon, WA, through northern WA, northern NT, Qld and NSW, to Vic and south-eastern SA. **HABITAT AND HABITS** Occurs in and around rivers, streams, lagoons, billabongs and ponds. Adults often seen at rest on tree roots, reeds and floating vegetation, with wings held longitudinally above abdomen, or flying low over water's surface. Larvae aquatic.

John Tann (cc)

Blue Riverdamsel ■ *Pseudagrion microcephalum* TL 35mm

DESCRIPTION Adults blue and blackish, with abdomen bronzed black on upper surface of segments 2–4 and blue at tip. Males have blue face and thorax. Wings membranous, clear, with metallic bronze sheen above and small black patches towards tip of each. Larvae dark greyish and elongate, and grow to *c*. 22mm. **DISTRIBUTION** Throughout most of Australian mainland from around Carnarvon, WA, through northern WA, northern NT, Qld and eastern NSW, to around Vic border; less common towards south of range. More widely distributed in Asia and other parts of Australasia. **HABITAT AND HABITS** Inhabits slow-moving and still waterbodies, including rivers, streams, lagoons, billabongs and ponds. Adults perch close to water's surface or on floating vegetation, with wings held longitudinally above abdomen. Larvae aquatic.

Dianne Clarke

Lestidae (Spreadwings)

Wandering Ringtail ■ *Austrolestes leda* TL 35mm; WS 45mm

Peter Rowland/kapeimages.com.au

Peter Rowland/kapeimages.com.au

TOP Male; BOTTOM Female.

DESCRIPTION Adults black and blue. Male brighter than female, with single blue ring on tip of abdomen and longitudinal blue mark, surrounded by black, on second abdominal segment. Immature female more brown. Wings membranous with black patches towards outer edge of each. **DISTRIBUTION** Throughout eastern Australia, including Tas, from NT–Qld border, through Qld, NSW and Vic, to south-eastern SA. **HABITAT AND HABITS** Found in and around a range of slow-flowing or still waterbodies, including swamps, ponds and rivers, although adults can disperse long distances away from these. Often seen perched horizontally on grasses or finer stems of vegetation with wings held along length of abdomen. Female lands on floating vegetation to lay eggs on water's surface, or can be still attached to male perched on protruding aquatic vegetation. Larvae aquatic, often seen on submerged vegetation.

MEGAPODAGRIONIDAE (FLATWINGS)

Common Flatwing ■ *Austroargiolestes icteromelas* TL 43mm; WS 70mm

DESCRIPTION Adults blackish with paler markings on head, back and slender abdomen, particularly towards tip. Wings membranous, clear, with metallic bronze sheen above and small black patches towards tip of each. Two subspecies: *A. i. icteromelas* has pale yellowish-white mouth; *A. i. nigrolabiatus* has dark brownish-black mouth. Larvae yellowish-cream and green with broadly elongated caudal gills at tip of abdomen, and growing to *c.* 26mm. **DISTRIBUTION** Endemic to eastern Australia, from around NT–Qld border, through Qld and eastern NSW, to western Vic. **HABITAT AND HABITS** Occurs around slow-moving rivers and streams, and less commonly around still pools and ponds. Adults perch on leaves with wings often held outstretched, but female sometimes holds wings longitudinally above abdomen. Adults feed on other flying insects, while aquatic larvae prey on mosquito larvae and other aquatic invertebrates.

Andrew Allen

PLATYCNEMIDIDAE (THREADTAILS OR WHITE-LEGGED DAMSELFLIES)

Orange Threadtail ■ *Nososticta solida* TL 35mm

DESCRIPTION Adult males largely black with orange markings on thorax, and large eyes. Abdomen has thin whitish lines between segments and white tip. Females similarly marked but pale brown. Wings yellowish (darker at bases in males) and membranous, held along length of abdomen when at rest. Larvae yellowish-brown and elongated, growing to *c.* 18mm. **DISTRIBUTION** Endemic to Australia; found throughout eastern parts, from northern Qld, through NSW, to Vic, and possibly to SA, where there is a single unconfirmed record. **HABITAT AND HABITS** Inhabits rivers, streams, lakes and dams, as well as small, stagnant pools. Adult males protect an area, defending it against would-be rivals with skillful flying ability. Once mated, females make slits in plant stems below water's surface to insert eggs, with or without male still attached. Adults frequently rest on vegetation near the water's edge.

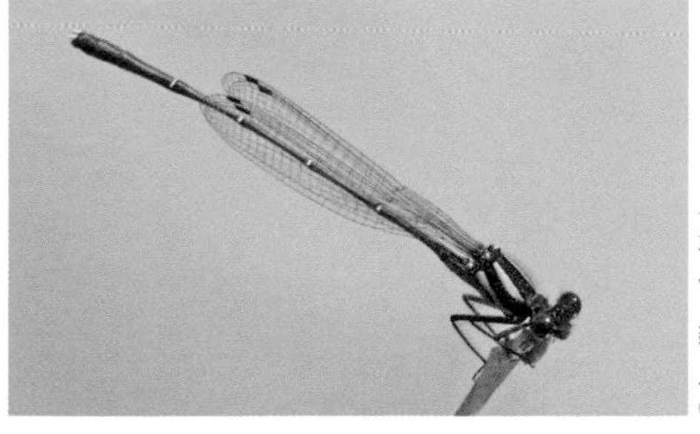

Graham Winterflood (cc)

PHASMATIDAE (STICK INSECTS)

Peter Street

Titan Stick Insect

■ *Acrophylla titan* TL 135–260mm

DESCRIPTION Greyish-brown with lighter pinkish blotches, dark conical tubercles on mesonotum, serrated edges to legs and long cerci. Hindwings (when exposed) have dark mottled pattern. Female much larger than male. Young nymphs greenish. **DISTRIBUTION** Eastern Australia, from south-eastern Qld to eastern Vic. **HABITAT AND HABITS** Occurs in forests, woodland and gardens. Feeds on leaves of a variety of trees, including tea trees, native cypress, wattles and eucalypts. Smaller males can fly and may be attracted to lights at night. Females cannot fly but still use their wings in a defence display when disturbed. One of the world's longest stick insect species.

Scott Eipper

Female and male (mating)

Tessellated Stick Insect

■ *Anchiale austrotessulata* TL 150mm

DESCRIPTION Medium-sized, stick-like insect. Thick body varyies in colour from light grey-brown to greenish; spines on legs; small, sparse spines on thorax; light-and-dark tessellated pattern on wings. Males smaller than females, with lighter build, wings with white stripe on leading edge, and longer antennae. Males' wings cover two-thirds of abdomen, while short wings of females only cover one-third. **DISTRIBUTION** Eastern Qld and NSW. **HABITAT AND HABITS** Most common in south-east Qld and coastal highlands of NSW, in forested areas such as eucalypt woodland, parks and gardens. Feeds on a variety of native trees, but typically on gum and wattle leaves. Large populations may defoliate entire trees. Adult males quite active and may fly into houses, attracted to lights at night.

Margin-winged Stick Insect ■ *Ctenomorpha marginipennis* TL 120–200mm

DESCRIPTION Normally grey to dark brown, although female sometimes dark green, with numerous raised tubercles on thorax. Male smaller than female with large brown wings; wings of female smaller, blackish and with distinctive tessellated rear margin. **DISTRIBUTION** South-eastern Australia, from southern Qld, through eastern NSW, Vic and Tas, to eastern SA. **HABITAT AND HABITS** Found in woodland and heathland. Feeds on leaves of various plants, including acacias, eucalypts and peas. Cryptic colouration provides excellent camouflage. Males are strong fliers, while females try to avoid predation by holding open their wings.

Dianne Clarke

Spur-legged Stick Insect

■ *Didymuria violescens* TL 105–110mm (Violet-winged Stick Insect)

DESCRIPTION Variable. Male typically brown, but can be green, with brown to purplish hindwings. Female typically green, but can be brown, with whitish to pink hindwings. Female larger and more robust than male, and male has 3 conspicuous 'spurs' on rear limbs. **DISTRIBUTION** Southern Australia and probably more widespread than records indicate. Most records from south-eastern Australia, from south-eastern Qld, through eastern NSW, to central Vic and Tas. **HABITAT AND HABITS** Occurs in a wide variety of wooded habitats, typically mountain forests. Feeds on gum leaves, particularly Mountain Ash *Eucalyptus regnans* and Alpine Ash *E. delegatensis*. Occasionally populations grow so large that they defoliate entire trees. While male is capable of flight, disturbed individuals typically drop to the ground and remain motionless, feigning death.

Peter Rowland/kapeimages.com.au

Ryan Francis

Goliath Stick Insect

■ *Eurycnema goliath* TL 200–250mm

DESCRIPTION Large stick insect. Adults bright green with yellow-and-white bands; wings have red undersides; spined legs. Adult female larger with thick abdomen; male slimmer with long wings and antennae. Nymphs brown and stick-like. **DISTRIBUTION** Northern and eastern Australia. **HABITAT AND HABITS** Occurs in eucalypt woodland and other forested areas, including parks and gardens. Feeds primarily on leaves of *Eucalyptus*, *Callistemon* and *Acacia*. If threatened, spreads wings to reveal red patches on undersides, and can make a hissing noise. Can also kick with thick spines on hindlegs in an attempt to deter predators. One of the world's largest stick insect species.

Spiny Leaf Stick Insect ■ *Extatosoma tiaratum* TL 130–200mm

(Macleay's Spectre)

Lorraine Harris

DESCRIPTION Medium-sized, dark brown to yellowish stick insect resembling a dead leaf. Spined; abdomen typically curled over body. Black-and-green lichen-mimicking form also occurs. Adult female larger, thicker with small wings; male slender with long wings and antennae. Young nymphs ant-like. **DISTRIBUTION** Eastern Australia. **HABITAT AND HABITS** Occurs in eucalypt woodland and rainforests, though can be found in parks and gardens where host trees occur. Lichen form occurs in high-elevation forest. Feeds mainly on gum leaves, but has been seen to feed on a wider variety of plants, including wattle and even rose. When threatened, can strike with spines on its legs, and capable of emitting a defensive odour similar to vinegar or toffee. Resembles a leaf, but is not a true leaf insect.

Large Pink-winged Stick Insect ■ *Podacanthus typhon* TL 108–145mm

(Pink-winged Phasma)

DESCRIPTION Mostly green insect, with narrow pronotum, reddish dorsal surface of abdomen and pink wings. Adult female resembles gum leaves. **DISTRIBUTION** Southern and eastern Australia. **HABITAT AND HABITS** Occurs in open woodland areas. Feeds mainly on *Eucalyptus* leaves. When threatened, opens its brightly coloured wings in a defence display.

Peter Rowland/kapeimages.com.au

Children's Stick Insect ■ *Tropidoderus childrenii* TL 200mm

DESCRIPTION Adult female predominantly green, resembling gum leaves; wings have yellow-and-blue patches at bases. Male slender, brownish-yellow, with white stripes on long wings. **DISTRIBUTION** Eastern Australia. **HABITAT AND HABITS** Found on trees in woodland areas, where it feeds predominantly on *Eucalyptus* leaves. When threatened, opens wings to reveal bright patches in a defence display. Female is an excellent mimic of gum leaves, but species is not a true leaf insect.

Sharon McGrigor

Graham Wise (cc)

Female. LEFT with wings extended; RIGHT with wings concealed

Phylliidae (Leaf Insects)

Monteith's Leaf Insect

■ *Phylium (Walaphylium) monteithi* TL 100mm

DESCRIPTION Pale yellow to bright green, with both sexes leaf-like. Female broader, male slimmer, with long, translucent wings. **DISTRIBUTION** Tropical north Qld. **HABITAT AND HABITS** Rarely encountered, it occurs in tropical rainforests. Feeds on leaves of rainforest plants in Myrtaceae family, such as the Golden Penda *Xanthostemon chrysanthus*, Cape Ironwood *Gossia floribunda* and some *Eucalyptus* spp. The only species of true leaf insect in Australia.

Per Thyme

Female and male (mating)

Calamoceratidae (Caddisflies)

Bicolour Sleeping-bag Caddisfly ■ *Anisocentropus bicoloratus* TL *c.* 10mm

DESCRIPTION Adults brown-orange with end half of wings black. Larvae unknown. **DISTRIBUTION** South-east Australia from south-eastern Qld, through eastern NSW, to western Vic. **HABITAT AND HABITS** Occurs around cool streams, and backwaters of large rivers or lakes. Pupae often found attached to rocks with silk. Larvae of other species in genus are detritivores, and their pupal case is flat and made of 2 pieces of leaf, hence the name 'sleeping-bag caddisflies'.

Donald Hobern (cc)

HYDROPSYCHIDAE (NET-SPINNING CADDISFLIES)

Net-spinning caddisflies ■ Multiple species TL up to 40mm; WS up to 40mm

DESCRIPTION Adults range from mottled brown to mostly white. Larvae usually large (to 20mm), curved and have branched gills along abdomen. **DISTRIBUTION** Worldwide, and suitable habitats in each Australian state and territory. **HABITAT AND HABITS** Inhabits flowing freshwater systems such as rivers and streams. Larvae do not build a typical case but construct retreats on sides of rocks, where they spin a silk mesh 'net' to catch algae, detritus and small invertebrates. Once fully grown, larvae undergo complete metamorphosis and pupate into adults.

Donald Hines (cc)

LEPTOCERIDAE (LONG-HORNED CADDISFLIES)

Long-horned caddisflies

■ Multiple species TL <20mm; WS 10–40mm

DESCRIPTION Adults almost black to light brown, and easily distinguished from other caddisflies and similar sized moths by their very long, thin antennae, as well as long palps that may appear as an extra pair of legs under head. **DISTRIBUTION** Worldwide, and recorded in suitable habitats in each Australian state and territory. **HABITAT AND HABITS** Typically occurs near fresh water, from large rivers to small ponds, though some species can tolerate some salinity. Larvae construct cases out of materials in the environment, such as vegetable matter or sand grains. They feed on algae and similar plant material, but can also be predatory.

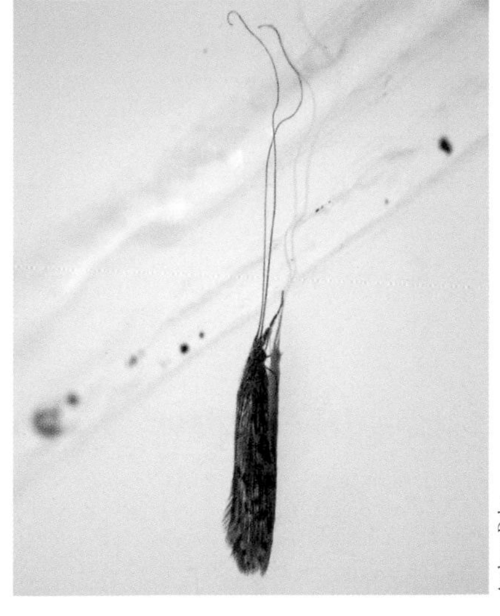

Anthony Daley

AMORPHOSCELIDAE (BOXER BARK MANTISES)

Bark mantises ■ *Gyromantis* spp. BL 25mm

DESCRIPTION Small, somewhat flattened mantis. Mottled light grey to brown with dark patches; pink patch on inner forelimb; abdomen segments ridged; short spines on head and pronotum. Adult female larger than male, with smaller wings. **DISTRIBUTION** North-western and eastern Australia. **HABITAT AND HABITS** Commonly found hunting small insects on trunks of trees in woodland areas, where it is very well camouflaged against the bark. Only adult males capable of flight. Females lay long, narrow ootheca attached to tree bark.

Lorenzo Bertola

Boxer bark mantises ■ *Paraoxypilus* spp. BL 20–30mm

DESCRIPTION Small mantis. Mottled brown to grey; orange-pink patches on forelimbs; femur of forelimb thickened; short spines above eyes. Adult male slimmer than female, with long, mottled wings. Section of raptorial forelimb thicker than in other mantids, almost oval in shape, giving forelimbs the appearance of clubs, or 'boxing gloves'. **DISTRIBUTION** Throughout Australia. **HABITAT AND HABITS** Occurs in parks, forests and other wooded areas, hunting small insects on trees and in leaf litter, where it is well camouflaged. Some species ant-like in appearance. Males capable of flight and may be attracted to lights at night. Females lay ootheca on trunks of trees.

Simon Grove

Liturgusidae (Bark & Lichen Mantises)

Tree runner mantises

Ciulfina spp. BL 24–38mm

DESCRIPTION Flattened body; mottled cream-and-brown pattern; long legs held out sideways from body; forelimbs held under body. Female has very short wings; male has longer wings covering more than half of abdomen. **DISTRIBUTION** Coastal Qld and northern Australia. **HABITAT AND HABITS** Occurs in a variety of wooded habitats, including rainforests, eucalypt woodland, mangroves and urban areas. Found on tree trunks, its colouration giving it excellent camouflage against the bark. Fast mover, actively hunting arthropod prey and running to avoid predators.

Rachel Whitlock

Mantidae (Praying Mantises)

Stick Mantid

Archimantis latistyla BL 90mm

(Brown Mantid)

DESCRIPTION Large, slender mantis. Pale grey to pale brown, sometimes with blue marking on head between eyes. Adult female larger with wings covering half of abdomen; male smaller with wings reaching tip of abdomen. Nymphs may be greener and can have white stripe down back. **DISTRIBUTION** Mainland Australia. **HABITAT AND HABITS** Occupies various habitats including gardens, but typically occurs in relatively dry regions near coasts. Hunts in grass, shrubs, wattles and small gum trees, predating on a wide variety of arthropods smaller than its body size. Often seen ambushing insects attracted to artificial lights. May be mistaken for a stick insect, but easily distinguished by its triangular head and raptorial forelimbs. Females lay large, brown, spongy, roughly squash ball-sized ootheca on branches of shrubs.

Andrew Allen

Giant Rainforest Mantid ■ *Hierodula (Hierodula) majuscula* BL 70–100mm

DESCRIPTION Large mantis. Body green with pale underside; pink-red tinge to legs; black markings on inner forelimbs; full wings to tip of abdomen in both sexes. Female larger than male, with 6 abdominal segments, while male has 7. **DISTRIBUTION** Tropical north-eastern Qld. **HABITAT AND HABITS** Occurs in tropical forests. Generalist predator hunting mostly arthropods such as grasshoppers, butterflies and dragonflies. Occasionally may capture and eat small vertebrates like frogs or geckos. Females lay large, pale, spongy ootheca on branches or foliage (pictured). One of Australia's largest mantids.

Donald Hobern (cc)

Green Mantid ■ *Orthodera ministralis* BL 40mm

(Garden Mantid)

DESCRIPTION Bright green mantis with yellow margins, broad, shield-like pronotum, blue spots on forelimbs, and full wings to tip of abdomen in adults of both sexes. Can have pinkish tinge to head, antennae and legs. **DISTRIBUTION** Throughout Australia. **HABITAT AND HABITS** Commonly seen in gardens and parks, and occurs in various habitats from open woodland to rainforests. Ambush predator, camouflaging itself in leafy bushes and shrubs, and targeting any small arthropod. Often spotted ambushing other insects drawn to artificial lights. Ootheca is brown with an almost woody appearance, and may be found on stems, leaves or man-made structures.

Peter Rowland/kapeimages.com.au

False Garden Mantid ■ *Pseudomantis albofimbriata* BL 60mm

DESCRIPTION Slender body; both green and brown colour forms; pale stripes on sides of head; pale white edges on pronotum and abdomen; dark spot on forelimb. Adult male has wings to tip of abdomen; female larger, with wings covering half of abdomen. **DISTRIBUTION** Eastern Australia. **HABITAT AND HABITS** Typically occurs in forests and woodland, usually on bushes, grasses and other plants, where it ambushes its arthropod prey. Similar to the Green Mantid (see opposite), it is also one of the most commonly encountered mantids in gardens and urban areas. Can be differentiated by its body being longer and slimmer, especially the pronotum.

Donald Hobern (cc)

Purple-winged Mantid ■ *Tenodera australasiae* BL 80mm

DESCRIPTION Slender mantid; body light brown to greenish in colour with pale edges; front half of eyes green. Green on leading edge of forewing; purple patches at bases of hindwings; full wings to tip of abdomen in both sexes. **DISTRIBUTION** Eastern Australia. **HABITAT AND HABITS** Commonly found in parks and woodland areas on grass, shrubs, wattle and small gum trees. Ambushes prey typically by waiting upside-down on grass or a stem, head facing the ground, ready to drop on prey passing below. Feeds on flying insects such as bees and flies, as well as other arthropods, and occasionally small vertebrates such as frogs. Females lays pale brown ootheca on leaves or branches. May be mistaken for a stick insect, but easily distinguished by its triangular head and raptorial forelimbs.

Kristi Ellingsen

Female and male (mating)

LEPTOPHLEBIIDAE (MAYFLIES)

Mayflies ■ *Atalophlebia* spp. WS 25mm

DESCRIPTION Adults dark bodied with 2 pairs of membranous wings; hindwings smaller. Legs usually banded, and 3 long appendages at end of abdomen. Male often has white tips to cerci. Nymphs wingless, with feathery gills on sides of abdomen. **DISTRIBUTION** Throughout Australia (except WA), including Tas. **HABITAT AND HABITS** Found near still and flowing waters. Adults short-lived and usually fly in swarms around waterbodies. Nymphs aquatic, occurring in clean fresh water, where they feed on vegetable matter and detritus.

Kristi Ellingsen

EUSTHENIIDAE (EUSTHENID STONEFLIES)

Eusthenid Stonefly

■ *Eusthenia spectabilis* TL >50mm

DESCRIPTION Body dark green-black. Adults winged, with white bands on wings when folded; bright orange colour visible when unfolded. Nymphs wingless, though they develop skin flaps on back, with obvious paired gills on abdomen. **DISTRIBUTION** Tas. **HABITAT AND HABITS** Typically occurs in and around fast-moving creeks and rivers, sometimes in lakes, with adults seen on boulders or in vegetation. Nymphs aquatic, generally clinging to rocks and other debris, but also capable of swimming. Unlike other stonefly species, nymphs are active predators of other invertebrates.

Kristi Ellingsen

Gripopterygidae (Gripopterygid Stoneflies)

Gripopterygid stoneflies ■ Multiple species TL 5–25mm

DESCRIPTION Body usually yellow to dark grey-brown or reddish. Nymphs have characteristic tuft of gills and 2 longer filaments at end of abdomen. Adults winged. **DISTRIBUTION** Southern Australia. **HABITAT AND HABITS** Typically inhabits fast-flowing waters, with a couple of species found in ponds and dams. Nymphs usually found under rocks and clinging to debris. Nymphs and adults feed mostly on plant matter and detritus.

Reiner Richter

Blaberidae (Blaberid Cockroaches)

Common Burrowing Cockroach ■ *Geoscapheus dilatatus* TL c. 75mm

DESCRIPTION Broad, rounded, wingless, dark reddish-brown cockroach, with legs covered in stout spines. **DISTRIBUTION** South-eastern Australia. **HABITAT AND HABITS** Occurs in semi-arid regions. Constructs a permanent underground burrow in sandy soil, coming up to feed on leaf litter at night. Most often seen at certain times of year, usually before or after rain, when large numbers emerge from their burrows and move across fields and highways en masse

Donald Hobern (cc)

Bark cockroaches ■ *Laxta* spp. TL 30mm
(Flat Cockroach, Trilobite Cockroach)

DESCRIPTION Flattened, oval-shaped, armoured appearance with camouflage colouration of mottled browns and greys. Adult male slender and usually winged; wings can be long or short depending on species. Adult female and nymphs rounded and wingless. **DISTRIBUTION** Eastern Australia. **HABITAT AND HABITS** Hides during the day in crevices or under bark, usually forming groups. Emerges at night to feed on the ground or on low vegetation. Winged males sometimes attracted to lights at night.

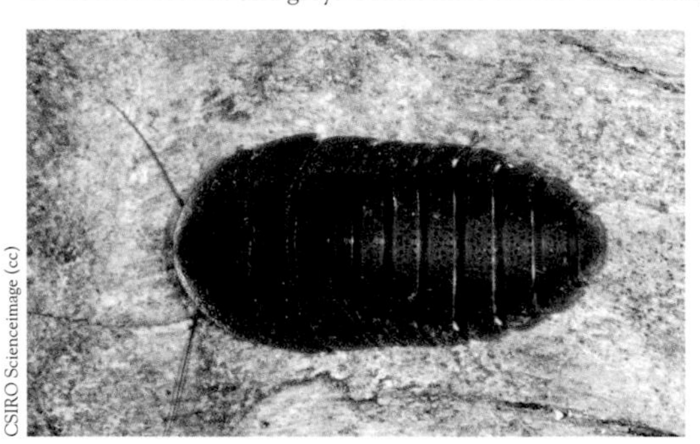
CSIRO Scienceimage (cc)

Giant Burrowing Cockroach ■ *Macropanesthia rhinoceros* TL 80mm
(Rhino Cockroach, Litter Bug)

DESCRIPTION Very large, wingless, reddish-brown cockroach, almost beetle-like. Legs covered in spines; forelegs spade-like. Pronotum concave, shovel-like. **DISTRIBUTION** North-eastern Qld. **HABITAT AND HABITS** Found in arid to semi-arid regions along coasts. Constructs permanent spiralling burrow to 1m deep x 6m long. Emerges periodically to restock burrow with supply of dry leaves that it eats. Males more commonly seen than females as they wander above ground at certain times of year, usually at the start of the rainy season. If disturbed, can make hissing noise in an attempt to deter predators. Australia's largest species of cockroach, and the world's heaviest species.

Rachel Whitlock

Speckled Cockroach ■ *Nauphoeta cinerea* TL 30mm

(Speckled Feeder Roach, Lobster Cockroach, Wood Cockroach, Woodies)

DESCRIPTION Light and dark brown mottled pattern. Adults have wings that almost or fully cover abdomen. Nymphs typically uniform shiny dark brown. **DISTRIBUTION** Worldwide, and native to north-eastern Africa. **HABITAT AND HABITS** Introduced to Australia (and many other countries). Commercially bred and sold as bait or feeder insects for pets like reptiles and birds, usually under various names such as 'feeder cockroaches' or 'woodies'. Escaped or released cockroaches may establish local feral populations but do not tend to thrive away from human habitation.

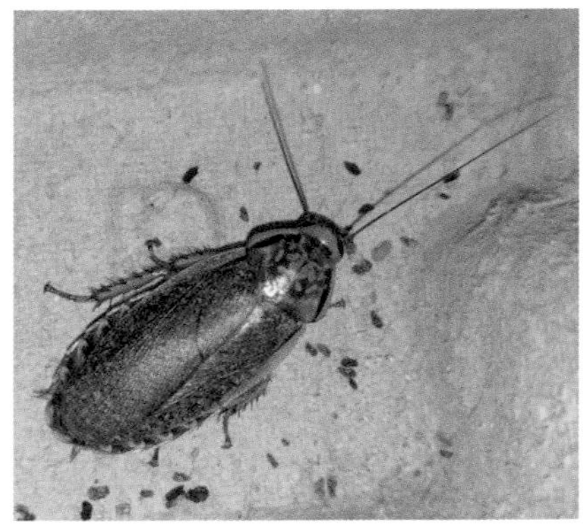

Rachel Whitlock

Australian Wood Cockroach ■ *Panesthia cribrata* TL 40mm

DESCRIPTION Broad, shiny black to dark brown, with spiny legs. Newly moulted adults have wings, which they soon bite or tear off, leaving stubs. **DISTRIBUTION** South-eastern Australia, Tas and Norfolk Island. **HABITAT AND HABITS** Found in rainforests and open forests. Lives in family groups inside rotten logs, which it burrows into. Also feeds on rotten logs and other decaying wood.

Auckland Museum (cc)

Surinam Cockroach ■ *Pycnoscelus surinamensis* TL 18–25mm

DESCRIPTION Wingless females most common, with shiny black body segments and last segments of abdomen duller in colour. Winged females and males have pale brown front edge to thorax and brown wings, and are poor fliers. **DISTRIBUTION** Native to Southeast Asia, and introduced to tropical and subtropical regions worldwide. **HABITAT AND HABITS** Commonly found burrowing in moist garden soil, pot plants, compost heaps, leaf litter and debris around human habitation, where it feeds on organic matter. Occasionally found inside homes. Sometimes a pest of crops, gardens and greenhouses, feeding on soft parts of plants. Known vector of poultry eye worm.

Robert Thacker

Blattidae (Blattid Cockroaches)

Native cockroaches ■ *Drymaplaneta* spp. TL 40mm

DESCRIPTION Medium to large cockroach. Black or dark brown body, some species having pale edges to sides of thorax. Hindwings absent. **DISTRIBUTION** Native to southern Australia. Two species introduced to NZ. **HABITAT AND HABITS** Shelters under bark or wood during the day, coming out to forage among leaf litter at night. May enter houses but do not establish themselves as pests. *D. semivitta* and *D. communis* are the most commonly encountered species.

Andrew Allen

Common Methana ■ *Methana marginalis* TL 25–30mm
(Methana Cockroach)

DESCRIPTION Typically dark brown to reddish-brown, with conspicuous broad cream edges along sides of body and wings (adult), extending to front of pronotum in a 'U' shape. **DISTRIBUTION** Native to eastern Qld and eastern NSW; introduced to Norfolk Island. **HABITAT AND HABITS** Terrestrial and arboreal, sheltering under bark and other vegetation in native bushland and urban gardens. Also enters houses.

Peter Rowland/kapeimages.com.au

American Cockroach ■ *Periplaneta americana* TL 55mm

DESCRIPTION Large reddish-brown cockroach, with pale yellowish-brown margin on pronotal shield. Large wings (extending past tip of abdomen in male) and a competent flier. **DISTRIBUTION** Native to Africa, and introduced to most countries around the world, including Australia. Occurs in close association with urban human habitation. **HABITAT AND HABITS** Mostly found indoors, where it is active at night. Pest species of urban households and similar dwellings. Favours darker areas during the day, including subfloor areas, roof voids, wall cavities and gardens. Eats most organic matter, particularly fermenting foods. Vector for a number of bacterial diseases and a leading cause of allergic reactions.

Peter Rowland/kapeimages.com.au

Australian Cockroach ■ *Periplaneta australasiae* TL 35mm

DESCRIPTION Brown, with yellow border around pronotum and longitudinal yellow stripes on outer edges of wings. Very similar to larger American Cockroach (see p. 37). **DISTRIBUTION** Introduced and now widespread in tropical and subtropical areas of Australia, in close proximity to human habitation. **HABITAT AND HABITS** Urban areas, where it is common in gardens and outside areas; also enters houses to find food, and is a competent flier. Pest species of households and similar dwellings, feeding on a variety of vegetable matter, but also scavenges on other organic matter. Known to spread disease-causing bacteria.

Rachel Whitlock

Mitchell's Diurnal Cockroach ■ *Polyzosteria mitchelli* TL 45mm

(Mardi Gras Cockroach)

DESCRIPTION Stout, wingless cockroach with striking colouration – metallic brown to blue-green body with yellow stripes, yellow-and-blue legs and yellow head. **DISTRIBUTION** Southern Australia. **HABITAT AND HABITS** Generally solitary cockroach occurring in semi-arid regions. Active during the day, usually on low vegetation. Also encountered on the ground at night. Can spray foul-smelling liquid from abdomen when disturbed.

CSIRO Scienceimage (cc)

Female carrying egg casing

Blattellidae (Blattellid Cockroaches)

German Cockroach ■ *Blatella germanica* TL 15mm

DESCRIPTION Pale brown to dark brown, with 2 dark longitudinal stripes on pronotum distinguishing it from all other similarly sized species. **DISTRIBUTION** Native to Africa, and introduced to Australia. **HABITAT AND HABITS** Closely associated with human habitation. Found most often around food preparation and storage areas in houses and commercial buildings, where it is largely nocturnal. Female carries ootheca within body. Identified in the spread of a number of diseases, including salmonella and typhus, and more than 170 bacterial isolates have been recorded on its body.

Imbuga (cc)

Pseudophyllodromiidae (Pseudophyllodromiid Cockroaches)

Balta cockroaches ■ *Balta* spp. 10–25mm

DESCRIPTION Most species vary from light to dark brown and black; some are intricately patterned. Adults usually winged, with translucent edges to wings and pronotum. **DISTRIBUTION** Throughout Australia. Torres Strait Islands. **HABITAT AND HABITS** Usually found on native vegetation at night, or in leaf litter during the day. Occasionally wanders into houses, sometimes attracted to lights at night. Some species may be mistaken for the German Cockroach (see above).

Kristi Ellingsen

Common Ellipsidion ■ *Ellipsidion humerale* TL 15mm
(Bush Cockroach)

Andrew Allen

DESCRIPTION Yellow-orange body with white bands; black and white underneath. Pale hind edge to pronotum. Adults winged; wings have intricate, net-like pattern. **DISTRIBUTION** Throughout Australia. **HABITAT AND HABITS** Found in bushland areas and sometimes gardens. Active on bushes and vegetation during the day. Omnivorous, and thought to feed on fungi, pollen and honeydew.

Austral Ellipsidion ■ *Ellipsidion australe* TL 20mm
(Bush Cockroach)

DESCRIPTION Dark orange-brown to black body with white bands. White-yellow edge around pronotum with dark orange to black centre. Adults winged; wings have intricate, pale, net-like pattern and dark tips. **DISTRIBUTION** Eastern and Western Australia. **HABITAT AND HABITS** Found in bushland areas and sometimes gardens. Forages on bushes and vegetation during the day. Omnivorous, and thought to feed on fungi, pollen and honeydew.

Andrew Allen

MASTOTERMITIDAE (TERMITES)

Giant Northern Termite ■ *Mastotermes darwiniensis* TL 13mm (soldier)
(Darwin Termite)

DESCRIPTION Soldiers reddish-brown, with round head and short, stout mandibles. Workers large and pale, with dark mandibles. Alates brown with mostly pale wings. **DISTRIBUTION** Northern Australia. **HABITAT AND HABITS** Nests occur in trunks and root crowns of trees and tree stumps, as well as underground. Long, enclosed galleries can extend as far as 200m, enabling colony to forage over a large area. Causes significant damage to timber structures such as fence posts, power poles and buildings. Also considered an agricultural pest as it can damage crops and animal products, and kill living trees by ringbarking. May also damage inedible substances such as rubber, metal and concrete in search of food.

CSIRO Scienceimage (cc)

RHINOTERMITIDAE (TERMITES)

Subterranean termites ■ *Coptotermes* spp. TL 6mm (soldier)

DESCRIPTION Soldiers have pale orange-brown, oval-shaped head, and long, thin, red-brown mandibles. Workers pale. Alates brown with pale brown wings. **DISTRIBUTION** Mainland Australia. Introduced to NZ. **HABITAT AND HABITS** Primary nest usually located underground or in base of tree, with galleries extending up to 100m from nest. In drier areas nest may be a mound. Secondary nests can be constructed in sites such as tree stumps, wooden poles and wall cavities. Species in this genus cause the majority of serious damage to buildings and timber in Australia. Soldiers aggressive and produce a milky fluid from the head when threatened.

Patrick Kavanagh (cc)

TERMITIDAE (TERMITES)

Magnetic Termite ■ *Amitermes meridionalis* TL 6mm (soldier)
(Compass Termite)

Peter Rowland/kapeimages.com.au

Distinctive Magnetic Termite mound

DESCRIPTION Soldiers have yellowish head and dark mandibles curved inwards, with single 'tooth'. Workers pale. Alates brown with pale wings. **DISTRIBUTION** Restricted to northern NT. **HABITAT AND HABITS** Found in grassland, where large eusocial colonies construct distinctive wedge-shaped mounds up to 4m tall. Each mound is orientated with its main axis facing north–south (giving the species its common name), so reducing its exposure to the full heat of the sun. Larger surfaces on mound receive sunlight only during early morning and late afternoon. Outer wall of mound is thickened and protects enclosed galleries that surround the almost solid core and house the termites. Worker termites collect dead grass from surrounding grassland, mainly at night, taking it to the mound, where it is chewed up and stored for the colony to feed on.

Rachel Whitlock

Peter Rowland/kapeimages.com.au

TOP Soldiers guarding workers; BOTTOM Termite mounds are used as nests by some bird species.

Arboreal Termite
■ *Nasutitermes walkeri* TL 7mm (soldier)
(Tree Termite)

DESCRIPTION Soldiers yellow-brown; dark, rounded head with single prominent 'spike'. Workers pale. Alates brown. **DISTRIBUTION** South-eastern Qld and eastern NSW. **HABITAT AND HABITS** Occurs in bushland along coasts. Primary nest established in root crown of a tree. Once the colony grows to a certain size arboreal nests are constructed higher up the tree and in other trees, on trunks or at bases of branches. Galleries running the length of the tree trunk connect nests with tunnels under the ground. Can damage timber, especially damp timber in contact with the ground, but not usually considered a significant pest.

Order Embioptera (Webspinners)

Webspinners ■ Several families, multiple species TL 4–15mm

DESCRIPTION Small, elongated, pale to glossy-black insect with swollen tarsi on front legs. Males may be winged. **DISTRIBUTION** Throughout Australia. **HABITAT AND HABITS** Constructs silken tube galleries with silk glands in 'boxing glove' tarsi to protect it and the rest of colony as it forages for food. Feeds on lichens, moss and other plant matter. Silken galleries may be encountered in many different environments on sheltered surfaces, such as under rocks, bark and leaf litter, and on tree trunks and fence posts. Winged adult males that may be seen are short-lived and do not feed.

Kristi Ellingsen

Acrididae (Grasshoppers and Locusts)

Giant Green Slantface ■ *Acrida conica* TL 70mm

(Longheaded Grasshopper)

DESCRIPTION Large, long, thin grasshopper; body bright green to pale yellow or light brown. Pinkish on top of abdomen, sometimes with pale stripes along body and wings; thin hindlegs very long. Elongate, conical head with eyes and flat antennae at top; pale brown stripe down sides. Nymphs brown to green. **DISTRIBUTION** Mainland Australia. **HABITAT AND HABITS** Occurs in open wooded areas and grassland. Feeds mostly on long grasses, where its body shape and colouration camouflage it well. Can also feed on various other plants, sometimes causing damage to gardens or crops. Despite its particularly long hindlegs it is a relatively weak jumper. Adults are weak fliers, often making a buzzing noise in flight. May be attracted to artificial lights at night.

Andrew Allen

Spur-throated locusts ■ *Austracris* spp. TL 35–65mm

DESCRIPTION Large adults light brown to grey, with contrasting black longitudinal stripes and mottling on forewings and clear hindwings. Long, full wings extend past tip of abdomen when folded. Spur on throat visible in some species; hindlegs spined. Nymphs vary from bright green to orangey-brown with black markings. **DISTRIBUTION** Throughout Australia. **HABITAT AND HABITS** Occurs in open woodland, grassland and recently cleared areas. Considered a pest due to its ability to form large swarms that reach plague proportions as they feed on pastures and crops such as sorghum and soybean.

Canley (cc)

Gumleaf Hopper ■ *Goniaea australasiae* TL 30–50mm

(Dead Leaf Grasshopper)

DESCRIPTION Medium to large grasshopper. Occurs in various colour forms, from light grey to orange-brown, mottled with fine dark spots. Distinctly arched central ridge on thorax. Adults have full wings extending past tip of abdomen. **DISTRIBUTION** Throughout Australia, particularly southern and eastern regions. **HABITAT AND HABITS** Inhabits various open habitats, including grassland and heaths. Usually seen on the ground during the day. Camouflages itself well among dead leaves as it rests. Most active at night, ascending plants such as wattles and eucalypts to eat the leaves. Similar to the closely related **Slender Gumleaf Hopper** G. *vocans*, but generally smaller, with more pronounced arch in thorax, and typically with solid-coloured antennae.

Kristi Ellingsen

Wingless Grasshopper ■ *Phaulacridium vittatum* TL 18mm

DESCRIPTION Small to medium grasshopper. Body grey to brown, occasionally with white lateral stripes down back. Legs generally paler in colour, with black mark halfway along inner sides of thick femurs of hindlegs. Adults occur in short-winged and long-winged forms. **DISTRIBUTION** Coastal Southern Australia. **HABITAT AND HABITS** Occurs in forested areas of higher rainfall, usually near coasts. Does not typically eat grasses, instead preferring leaves of herbs, shrubs and some trees. Large populations can become pests in gardens, pastures and orchards. Has powerful hindlegs and is a strong jumper.

Peter Rowland/kapeimages.com.au

Giant Grasshopper ■ *Valanga irregularis* TL 50–80mm

(Hedge Grasshopper, Giant Valanga)

DESCRIPTION Large grasshopper. Light brown to grey body with dark spots and mottling, ridge down midline of thorax and throat spur. Hindwings usually grey, with 2 dark bands on femur and black-tipped orangey spines on tibia. Female larger than male. Nymphs variable in colour from bright yellow-green to brown with black, yellow, orange and pink markings. **DISTRIBUTION** Tropical and subtropical Australia. **HABITAT AND HABITS** Favours moist climates, occurring in grassland and forested areas. Feeds mostly on leafy shrubs such as native hibiscus. Often found in well-watered gardens on bushes and hedges. Has a throat peg like the spur-throated locusts (*Austracris* spp., see opposite) and may appear similar, but is much larger, usually with coloured hindwings and lacking distinct body stripes of the locusts, and is typically sedentary. Australia's largest grasshopper species.

Rachel Whitlock

PYRGOMORPHIDAE (GAUDY GRASSHOPPERS AND PYRGOMORPHS)

Rachel Whitlock

Vegetable Grasshopper

■ *Atractomorpha similis* TL 40mm (Common Grass Pyrgomorph, Northern Grass Pyrgomorph)

DESCRIPTION Compact bright green body, tapered at tip of head and abdomen, with pale yellow to white edges on head and thorax, and flattened antennae. Brown form also occurs. Dorsal surface of abdomen pinkish, hidden by folded wings. Hindlegs long and thin. **DISTRIBUTION** Throughout Australia. Also PNG. **HABITAT AND HABITS** Typically found in areas of high humidity and rainfall, often near water sources or near coasts. Feeds on leaves of herbs, broadleaved shrubs and low trees, and may become a pest in well-watered gardens. Hides in grass and relies on camouflage to avoid predators, preferring not to jump or fly away.

ANOSTOSTOMATIDAE (KING CRICKETS)

Giant King Cricket ■ *Anostostoma australasiae* TL 70mm

DESCRIPTION Adults wingless; body light to dark brown with reddish-brown head; long, thin antennae; long, spined legs. Adult male has large head with pronounced jaws.

Jordan de Jong

DISTRIBUTION South-east Qld and north-east NSW. **HABITAT AND HABITS** Found in rainforest habitat, where it constructs burrows in the soil. Omnivorous, emerging on wet nights to hunt for small arthropods and rotting fallen fruits. Closely related to the wetas of NZ.

GRYLLIDAE (TRUE CRICKETS)

House Cricket ■ *Acheta domesticus* TL 20mm

(Feeder Cricket)

DESCRIPTION Typically pale yellowish-brown with dark eyes and dark brown markings on thorax and abdomen, though body colour can vary to very dark brown and grey. Wings cover abdomen in adults. Superficially similar to the Tropical House Cricket (see below), but body not dorsally flattened and antennae wider apart at base. **DISTRIBUTION** Introduced to Australia. Worldwide, and probably native to south-western Asia. **HABITAT AND HABITS** Like in many other countries, commercially bred and sold as bait or as feeder insects for pets like reptiles and birds. Escaped or released crickets may establish local feral populations, but do not tend to thrive away from human habitation.

Brian Gratwick (cc)

Tropical House Cricket ■ *Gryllodes supplicans* TL 12–22mm

DESCRIPTION Pale yellowish-brown with single black transverse line between eyes. Body dorsoventrally flattened. Adult male has wings, used in production of song, and female has long ovipositor. **DISTRIBUTION** Introduced to Australia. Worldwide, and probably native to southern Asia. **HABITAT AND HABITS** Like in many other countries, may be sold as bait, as feeder insects for pets like reptiles and birds, or used for scientific research. Escaped or released crickets may establish local feral populations, but do not tend to thrive away from human habitation.

Peter Rowland/kapeimages.com.au

Black Field Cricket ■ *Teleogryllus commodus* TL 25–30mm

DESCRIPTION Dark brown to black, with wings ending in long point extending to rear of abdomen, long antennae and rear legs longer that other 2 pairs for jumping. Female has long, thin ovipositor. **DISTRIBUTION** Widespread in eastern and south-western Australia. Introduced into NZ. **HABITAT AND HABITS** Found in open forests, heaths, pasture, farmland and gardens. Lives concealed under fallen debris, in grasses, or in shallow burrows or soil cracks, emerging at night to feed on a variety of plant matter. More often heard than seen, with males singing repeatedly at dusk by rubbing their wings together. Often enters houses, attracted to lights at night.

Kristi Ellingsen

GRYLLOTALPIDAE (MOLE CRICKETS)

Southern Mole Cricket ■ *Gryllotalpa australis* TL 35mm

DESCRIPTION Elongate cricket with thickened head and enlarged first segment of thorax, small dark eyes, and dark brown to sandy orange-brown body. Front legs more robust than rear ones, with large claws for digging. Short, folded wings do not cover abdomen. Two long cerci at tip of cylindrical abdomen. **DISTRIBUTION** Southern NSW, Vic, SA and Tas. **HABITAT AND HABITS** Usually occurs in open, grassy woodland, as well as urban parks and gardens. Spends most of its life underground, tunnelling through soil and eating plant roots; occasionally seen above ground, particularly in wet weather. Males use their burrows to amplify their song. Omnivorous, and may emerge from burrow to forage for other foods including small arthropods. Capable of squirting a foul-smelling liquid if disturbed.

Simon Grove

Common Mole Cricket ■ *Gryllotalpa pluvialis* TL 45mm

DESCRIPTION Elongate cricket with thickened head and enlarged first segment of thorax, small dark eyes, and dark brown to sandy orange-brown body. Front legs more robust than rear ones, with large claws for digging. Short, folded wings do not cover abdomen. Two long cerci at tip of cylindrical abdomen. **DISTRIBUTION** Eastern Australia; introduced to WA. **HABITAT AND HABITS** Usually occurs in open, grassy woodland. Spends most of its life underground, tunnelling through soil and eating plant roots; occasionally seen above ground, particularly in wet weather. Omnivorous, and may emerge from burrow to forage for other foods including small arthropods. Males use their burrows to amplify their song, which is commonly heard in suburban gardens and parks at dusk or during rainfall, even if the insects themselves are not usually seen, though sometimes they can be attracted to lights. Capable of squirting a foul-smelling liquid if disturbed.

Jenny Thynne

Tettigoniidae (Katydids and Bush Crickets)

Mountain Katydid ■ *Acripeza reticulata* TL 50mm
(Mountain Bush Cricket)

DESCRIPTION Body grey to brown with darker mottling. Abdomen bright red with blue-and-black markings, usually covered by forewings. Adult female broad bodied with no hindwings. Adult male has full forewings and hindwings extending past tip of abdomen. **DISTRIBUTION** Eastern Australia. **HABITAT AND HABITS** Occurs in open forests on mountains, as well as in lowlands like paddocks. Usually found on the ground during the day, sometimes in large numbers, where it is camouflaged well against dead leaves. Feeds on leaves of various shrubs and herbaceous plants. Known for its defence display, during which it lifts its dull brown wings to reveal the brightly coloured abdomen, inflates a bright orange membrane behind the head and may produce droplets of fluid from membranes in the abdomen.

John Harris

Common Garden Katydid ■ *Caedicia simplex* TL 40mm

(Green Katydid)

DESCRIPTION Bright green body with long legs and long, leaf-like wings extending past tip of abdomen. Dorsal surface of abdomen has pink-and-yellow markings usually covered by wings. Nymphs can be a variety of colours, from green to pink and brown. Pink or yellow adults occasionally occur. **DISTRIBUTION** Native to Australia and NZ. **HABITAT AND HABITS** Occurs in wetter regions, typically near coasts, and most common in temperate areas of south-east Australia. Feeds on leaves and flowers of various plants, including cultivated plants in gardens. Males' soft, short call can often be heard in suburban areas in the evening.

Kristi Ellingsen

Thirty-two Spotted Katydid ■ *Ephippitytha trigintiduoguttata* TL 65mm

(Mottled Katydid)

DESCRIPTION Large green katydid with long wings, and dark brown to black spots around edges of forewings and on long legs. Early instar nymphs ant-like. **DISTRIBUTION** Eastern and northern Australia. **HABITAT AND HABITS** Widespread in open forests, feeding mainly on leaves in tops of gum trees. Pattern and shape of forewings superficially resemble a gum leaf with herbivory damage. Occasionally attracted to artificial lights at night.

Dianne Clarke

Spiny Rainforest Katydid ■ *Phricta spinosa* TL 60mm

(Prickly Katydid, Rainforest Tree Katydid, Giant Spiny Tree Cricket)

DESCRIPTION Body has intricate camouflage pattern of stripes, marbling and spotting in blue-green, pink, black and white, and is covered in spiny protrusions. Rests with all legs outstretched. **DISTRIBUTION** Tropical north Qld. **HABITAT AND HABITS** Native to tropical forests and rainforests of northern Qld. During the day, as it rests on tree trunks, its spines and colouration camouflage it well against lichens and mosses growing in the wet climate. Active at night, feeding on tree leaves and bark. During strong winds and storms, descends tree trunks to find a safe spot to hide; this is often when it is spotted. Capable of using large spines on long hindlegs for defence against predators.

Scott Eipper

Forficulidae (Earwigs)

European Earwig ■ *Forficula auricularia* TL 15mm

DESCRIPTION Dark reddish-brown body with pale brown legs and edges to pronotum. Adult male has smaller body, with larger and more strongly curved, toothed forceps, than female. **DISTRIBUTION** Introduced to Australia, NZ and North America. Native to Europe, western Asia and North Africa. **HABITAT AND HABITS** Prefers cool, moist environments, where it is active at night. Hides in leaf litter, crevices and other shelters during the day. Omnivorous, predating on other arthropods, and also feeds on fruits and other plant material; may occasionally become a pest, causing damage to agricultural crops.

Peter Rowland/kapeimages.com.au

Labiduridae (Earwigs)

Common Brown Earwig ■ *Labidura riparia* TL 35mm

DESCRIPTION Light brown with dark brown markings on thorax, wings and abdomen. Male has larger and more strongly curved forceps than female. **DISTRIBUTION** Coastal regions throughout Australia. **HABITAT AND HABITS** Common in various environments, particularly in sandy habitats. Prefers to shelter in dark, moist places during the day, emerging to hunt at night, when it also may be attracted to lights. Predates on other arthropods, preferring to feed on caterpillars, including ones larger than itself that it subdues with its forceps. These are also used in defence, wing folding and mating. Can emit a foul-smelling substance when disturbed.

Peter Rowland/kapeimages.com.au

Informal group Psocoptera (Booklice, Barklice and Psocids)

Webbing barklice ■ Family Archipsocidae, multiple species TL 1–4mm

Rachel Whitlock

DESCRIPTION Tiny insect, louse or termite-like in appearance. Mostly brown body, large dark brown head wider than thorax, and thin antennae. Adults have reduced wing veins. **DISTRIBUTION** Tropical and subtropical regions worldwide. **HABITAT AND HABITS** Typically found on bark of trees, feeding on fungi, algae and other microflora. Unlike many other psocids, species in this family usually occur in large groups under sheets of webbing that may cover entire tree trunks.

Booklice ■ *Liposcelis* spp. TL 2mm

DESCRIPTION Tiny insect. Flattened, louse-like appearance with longer, thin antennae, head larger than thorax and usually brown in colour. **DISTRIBUTION** Worldwide. **HABITAT AND HABITS** Feeds on organic matter, and many species are associated with human habitation. The cosmopolitan species *L. bostrychophila* and several others are common stored product pests, typically of damaged grain or grain contaminated with mould. Also known to feed on natural history collections in museums, paper and book-binding paste, hence the common name.

Anthony Daley

Barklice ■ Family Psocidae, multiple species TL 1–10mm

DESCRIPTION Tiny insect. Louse or termite-like, soft body, broad head, often humpbacked appearance and long, thin antennae; adults have long, tent-like wings. Colour and pattern varies, but typically a mix of brown, black and white. **DISTRIBUTION** Worldwide. **HABITAT AND HABITS** Commonly occurs on bark of trees, feeding on organic matter such as fungi, algae and other microflora. Usually sedentary or found in small groups. The psocid family includes the largest species of barklouse in Australia and the world. Sometimes mistaken for aphids or whiteflies, barklice can be distinguished by their large, broad heads.

Dianne Clarke

INFORMAL GROUP PTHIRAPTERA (PARASITIC LICE)

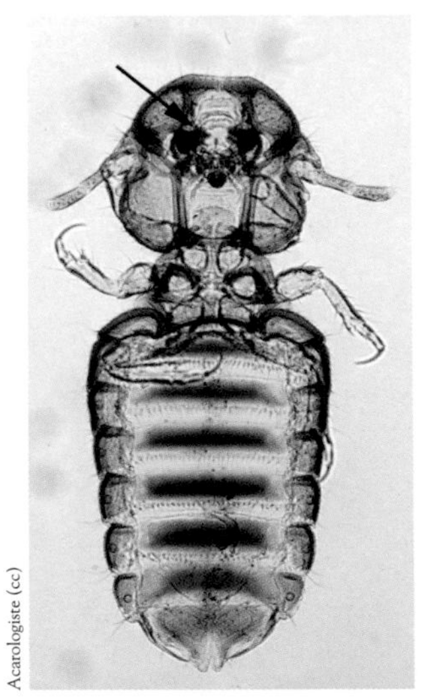

Acarologiste (cc)

Sheep Body Louse

■ *Bovicola (Bovicola) ovis* TL 2mm
(Sheep Louse)

DESCRIPTION Pale yellowish body, light brown, broad head with chewing mouthparts, and brown stripes on abdomen. **DISTRIBUTION** Worldwide, and introduced to Australia. **HABITAT AND HABITS** Introduced on its primary host, the sheep. Chewing louse that feeds on skin-gland secretions, skin cells and skin bacteria. Infestation can cause matting and discolouration of wool in sheep. Two other blood-sucking louse species also use sheep as primary hosts, the **Face Louse** *Linognathus ovillus* and **Foot Louse** *L. pedalis*, which have longer, thinner heads and appear more blue in colour.

British Natural History Museum (cc)

Horse Sucking Louse

■ *Haematopinus asini* TL 3mm

DESCRIPTION Greyish body, large abdomen and small, narrow head with sucking mouthparts. **DISTRIBUTION** Worldwide, and introduced to Australia **HABITAT AND HABITS** Primary hosts are horses and donkeys. Sucking louse, feeding on blood of host. Typically found in longer hair of mane, forelock and base of tail, although in severe infestations can be found all over the body. The **Chewing Louse** *Werneckiella equi* also feeds on horses and donkeys; it is reddish-brown in colour with a broad head.

Cattle Louse ■ *Linognathus vituli* TL 2.5mm
(Long-nosed Sucking Louse)

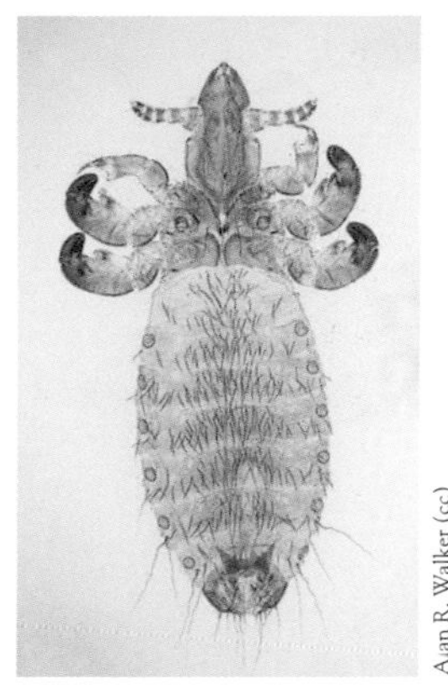
Alan R. Walker (cc)

DESCRIPTION Bluish-black body, elongated, narrow head with sucking mouthparts and large abdomen.
DISTRIBUTION Worldwide, and introduced to Australia.
HABITAT AND HABITS Sucking louse, feeding on blood of host, which is cattle. Can be found anywhere in a host's coat, such as folds of skin around neck and inner thighs. Heavy infestations can cause so much irritation to a cow that it can rub the hair off the skin, as well as causing conditions like anaemia. Other lice infest cattle, two common species being the **Chewing Louse** *Bovicola bovis*, which is reddish-brown with a broad head, and the **Short-nosed Sucking Louse** *Haematopinus eurysternus*, which is dark grey with a shorter head.

Poultry Body Louse ■ *Menacanthus stramineus* TL 3.5mm

DESCRIPTION Yellowish-brown body, and broad head with chewing mouthparts.
DISTRIBUTION Worldwide, and introduced to Australia **HABITAT AND HABITS** Primary hosts are poultry species, especially domestic chickens and turkeys. Chewing louse that feeds on skin-gland secretions, skin cells and skin bacteria. Can cause skin irritation and, in large infestations, conditions such as feather loss. Another louse species commonly infesting domestic poultry is the **Feather-shaft Louse** *Menopon gallinae*, a smaller chewing louse measuring 1.5–2mm long.

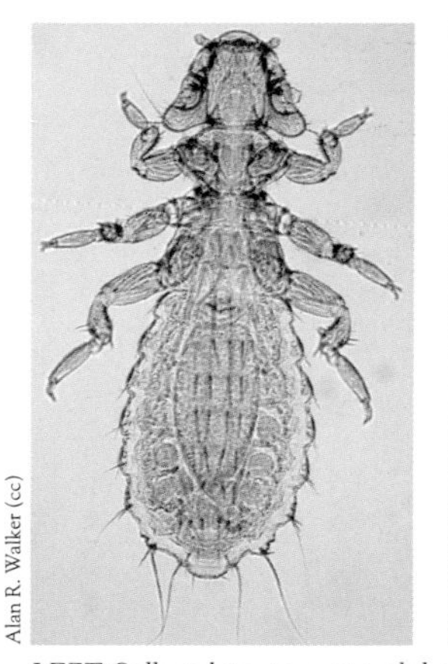
Alan R. Walker (cc)

John C. Abbott

LEFT Collected specimen viewed through microscope; RIGHT Live louse amongst feathers of host.

Human Body Louse & Head Louse

■ *Pediculus humanus* TL 4mm

DESCRIPTION Human Body Louse (subspecies *humanus*) and Head Louse (subspecies *capitis*). Wingless insects with large abdomen and slightly narrower head. Legs have sharp claws for gripping hair or clothing of host, and piercing and sucking mouthparts. **DISTRIBUTION** Worldwide, wherever humans congregate. **HABITAT AND HABITS** Lives on clothing of humans in crowded conditions with poor hygiene, and spread by close contact with other humans in same area. Feeds on blood, which can cause skin reactions including rashes and itchiness, and known to spread diseases such as epidemic typhus, trench fever and louse-borne relapsing fever. Does not persist in clean environments with regular bathing, and laundering of clothes and bed linen.

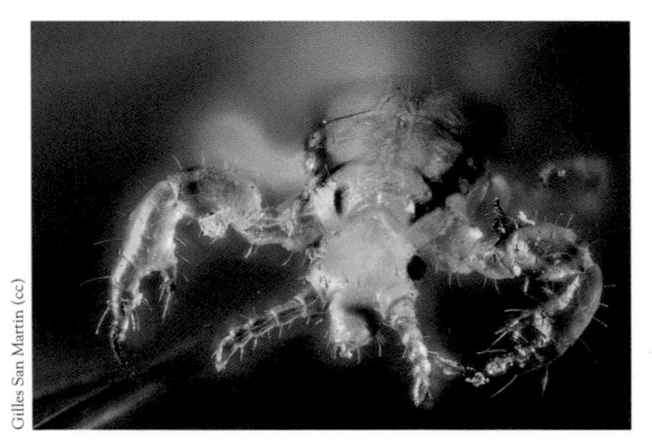

Gilles San Martin (cc)

Crab Louse ■ *Pthirus pubis* TL 2mm

(Pubic Louse)

DESCRIPTION Wingless grey insect with broad body and small head with sucking mouthparts. Legs have sharp claws for gripping hair or clothing of host. **DISTRIBUTION** Worldwide, wherever humans congregate. **HABITAT AND HABITS** Blood-sucking louse inhabiting parts of human body with coarse hair, such as armpits, thighs, groin and sometimes facial hair. Bites can cause itchiness and occasionally more severe skin conditions. Proliferates in unsanitary and overcrowded conditions, where it can easily move between hosts.

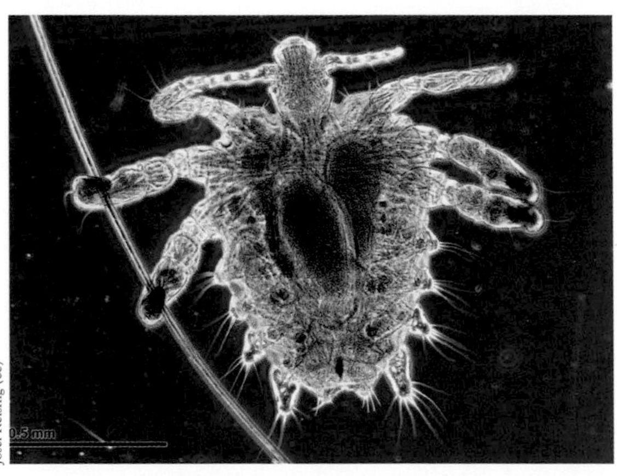

Josef Reishig (cc)

THRIPIDAE (THRIPS)

Greenhouse Thrips ■ *Heliothrips haemorrhoidalis* TL 1–2mm
(Glasshouse Thrips, Black Tea Thrips)

DESCRIPTION Adults winged, with dark brown bodies with paler tip to abdomen, and pale legs. **DISTRIBUTION** Native to South America, and introduced worldwide. **HABITAT AND HABITS** Found on a wide variety of plants, from tea plants to pine trees and ferns, where colonies congregate on undersides of leaves. Known for taking advantage of stressed or diseased plants. Very small insect, so its presence is usually detected by damage it causes to its host as it feeds, ranging from pale spots on leaves and brown patches on fruits, to leaf necrosis and defoliation. Considered an agricultural pest.

Jesse Rorabaugh (cc)

Eucalyptus Thrips ■ *Thrips australis* TL 1–2mm
(Gum Tree Flower Thrips)

DESCRIPTION Winged adults; colour variable from yellow-brown to pale near-white. **DISTRIBUTION** Native to eastern Australia, Lord Howe Island and Norfolk Island; introduced worldwide on *Eucalyptus* trees. **HABITAT AND HABITS** Typically found in flowers on *Eucalyptus* spp. (gum trees), as well as *Melaleuca* spp. (paperbarks), other Myrtaceae and *Acacia* spp. (wattles). Can be found in flowers of a wide range of nearby plants after host flowers die off. Feeds mostly on tree sap.

Laurence Mound

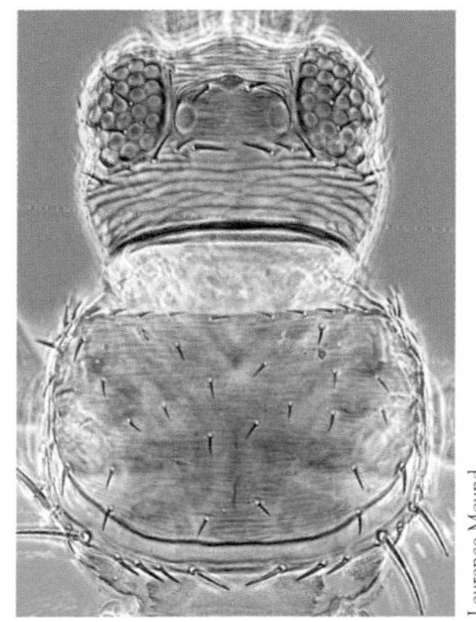

Laurence Mound

LEFT View of whole body through microscope; RIGHT Close-up of head and pronotum.

Phlaeothripidae (Tube-tailed and Giant Thrips)

Giant thrips ■ *Idolothrips* spp. TL 1–14mm

DESCRIPTION Large thrips easily seen with the naked eye. Adults winged, with dark brown-black, elongated body and head. Last abdominal segment elongated and tube shaped. **DISTRIBUTION** Eastern Australia, northern Australia and south-west WA. **HABITAT AND HABITS** Typically inhabits eucalypt woodland, and thought to feed on fungal spores on dead *Eucalyptus* leaves.

Kristi Ellingsen

Cercopidae (Spittlebugs)

Common Spittlebug ■ *Philagra parva* TL 8mm
(Froghopper)

DESCRIPTION Light orangey-brown body with darker markings. Dark horn- or nose-like projection curved upwards at tip of head. Adults fully winged. Nymphs paler, typically covered in white frothy mass. **DISTRIBUTION** Throughout Australia; common in eastern Australia. **HABITAT AND HABITS** Found on stems and branches of shady stands of vegetation. Feeds on sap of various native shrubs and small trees. Nymphs use excess honeydew to produce the distinctive mass of bubbles often found on plant stems, sometimes called 'cuckoo-spit', to avoid desiccation and predation.

Andrew Allen

CICADELLIDAE (LEAFHOPPERS)

Gum leafhoppers ■ *Eurymeloides* spp. TL 5–11mm

DESCRIPTION Small, cicada-like, wedge-shaped insects, varying in body colour but mostly black with bright yellow-orange and white markings. Adults fully winged. **DISTRIBUTION** Throughout Australia. Also PNG. **HABITAT AND HABITS** Occurs in a variety of habitats, predominantly eucalypt woodland. All life stages feed primarily on sap of *Eucalyptus* trees. Strong jumper, hence the name, and adults can fly.

Kristi Ellingsen

CICADIDAE (CICADAS)

Floury Baker ■ *Aleeta curvicosta* BL 50mm
(Floury Miller)

DESCRIPTION Stout cicada. Clear wings with two spots towards the tip of each forewing. Body dark brown-black with pale yellow-brown markings; entire body covered in fine white filaments, giving a powdery appearance. **DISTRIBUTION** Coastal eastern mainland Australia. **HABITAT AND HABITS** Typically occurs in a variety of habitats from mangroves and coastal heaths, to woodland and gardens. Uses piercing mouthparts to feed on xylem of various plants, preferring paperbarks *Melaleuca* spp. Adults commonly encountered from late November through to May. Only males call, sounding like a loud, rapid *zeep-zeep*, followed by a long hiss.

Dianne Clarke

Greengrocer ■ *Cyclochila australasiae* BL 40mm

(Yellow Monday, Masked Devil, Chocolate Soldier, Blue Moon, Red Warrior)

DESCRIPTION Various colour forms, including green, yellow, brown and turquoise. Wings vary from clear to pale blue; eyes usually red. **DISTRIBUTION** South-eastern Australia. **HABITAT AND HABITS** Occurs in wet and dry sclerophyll forests, usually near coasts, and feeds on plant sap. The variety of common names refers to the many colour forms that occur in different localities throughout the range. Adults commonly seen from September and through summer months. One of the loudest insects in the world, the call a mix of pulsing and continuous ringing.

Doug Beckers (cc)

Razor Grinder ■ *Henicopsaltria eydouxii* BL 55mm

DESCRIPTION Large cicada. Body dark brown with light brown margins and markings. Individuals have varying extents of light grey mottling. Wing membranes clear with dark and light wing veins. **DISTRIBUTION** Eastern mainland Australia **HABITAT AND HABITS** Found in wet and dry sclerophyll forests and rainforests near coasts, where it feeds on plant sap. Adults typically emerge in summer months of December and January, when large groups sing together. Call loud and abrasive, like the sound of a knife on a grindstone.

Dianne Clarke

Bark buzzers ■ *Pauropsalta* spp. BL 18–40mm
(Bark squeakers)

DESCRIPTION Body varies in colour from black with yellow margins, to grey and reddish-brown. Wings have dark veins and clear membranes. **DISTRIBUTION** Throughout Australia. **HABITAT AND HABITS** Occurs in a wide variety of wooded habitats, from arid zones to monsoonal forests, where it feeds on plant sap. Adults typically encountered during summer months. Song often complex, with components varying from a rapid *zip-zip* to a buzzing.

Kristi Ellingsen

Redeye ■ *Psaltoda moerens* BL 40mm
(Cherry Eye)

DESCRIPTION Dark brown to black body, grey-silver patches on sides of abdomen and striking bright red eyes. Wing veins black with clear membranes. **DISTRIBUTION** South-eastern Australia from south-east Qld to SA; also Tas. **HABITAT AND HABITS** Occurs in sclerophyll forests and open scrubland, feeding primarily on eucalypts, preferring smooth-barked species. In high population concentrations honeydew expelled from these insects in tall trees can seem like rainfall. Adults encountered in November–March, though population sizes vary from year to year. Song complex, a continuous rattling starting low and increasing in volume before breaking into a 'yodeling' sequence.

Peter Rowland/kapeimages.com.au

Brown Bunyip ■ *Tamasa tristigma* BL 20mm

DESCRIPTION Small cicada. Pale cream to brown body with light orange-brown and darker brown markings, black band near tip of abdomen and dark eyes. Wing veins pale brown with black spots near wing-tips; membranes clear. **DISTRIBUTION** Eastern Qld and NSW. **HABITAT AND HABITS** Occurs in open forests, where it typically feeds on sap of *Casuarina* spp. and wattle trees. Relies on colouration to camouflage itself against trunks of trees it feeds on. Adults usually sighted in November–April, and make a continuous buzzing call.

Donald Hobern (cc)

Double Drummer ■ *Thopha saccata* BL 50mm

DESCRIPTION Large, broad cicada. Body black with extensive yellow-orange to reddish-brown markings, particularly on pronotum and thorax. Males have 2 large, reddish-brown 'drums' on either side of abdomen. Wings have orange and dark brown veins, and clear membranes. **DISTRIBUTION** Eastern Australia, from around Cooktown, Qld, to Moruya, NSW. **HABITAT AND HABITS** Occurs in open sclerophyll forests, typically on high branches of gum trees, where it feeds on plant sap. Named after 2 reddish-brown pockets on either side of abdomen that amplify the males' call, which sounds like a continuous piercing whine that can change pitch. Adults usually spotted in November–March. Australia's largest cicada species, the first Australian cicada species to be described and one of the world's loudest insects.

John Nielsen

FLATIDAE (PLANTHOPPERS)

Green mottled flatid planthoppers ■ *Siphanta* spp. TL 10–20mm

(Green planthoppers, Torpedo bugs)

DESCRIPTION Small insect. Adults bright green to pale blue-green with fine pale spotting, sometimes with pale yellow or pinkish margins. Full wings give laterally compressed triangle- or wedge-shaped appearance. Nymphs more dorsally flattened, typically white or pale colour with darker spotting and white tuft at tip of abdomen. **DISTRIBUTION** Throughout Australia and surrounding islands. **HABITAT AND HABITS** Occurs in various habitats and capable of feeding on sap of a wide range of plants. Strong jumper, hence the name 'torpedo bugs' for the nymphs, and adults can fly. Most of the insects in this family look very similar and require specialist identification to tell species apart.

Rachel Whitlock

MEMBRACIDAE (PLANTHOPPERS)

Lantana Treehopper ■ *Aconophora compressa* TL 8mm

DESCRIPTION Adults have dark brown and reddish-brown body, large, single, thick spine protruding above head and thorax, and full, clear wings. Nymphs wingless, cream to light brown with dark stripes, covered with small spines. **DISTRIBUTION** Native to Mexico and introduced to Australia, where it occurs in south-east Qld and north-east NSW. **HABITAT AND HABITS** Introduced as a biological control agent for the invasive weed *Lantana* spp. Usually found in large congregations feeding on phloem in stems of *Lantana*. Can cause dieback of host plants in high enough population concentrations. Also occasionally feeds on other plants, typically in the same plant family as *Lantana*, such as the ornamental Fiddlewood Tree *Citharexylum spinosum*.

Rachel Whitlock

RICANIIDAE (PLANTHOPPERS)

Passionvine Hopper ■ *Scolypopa australis* TL 7mm

DESCRIPTION Small, cicada-like insect. Light brown body with white and dark markings. Adults have large, broad wings with brown and clear patches. Nymphs wingless, with thick white tuft of filaments at tip of abdomen. **DISTRIBUTION** Eastern Australia. Introduced to NZ. **HABITAT AND HABITS** Occurs in temperate areas. Feeds on sap of various ornamental, introduced and native plants, such as passionfruit vines, kiwifruit and *Lantana*. All life stages are strong jumpers, and adults are capable of flight. Bees as well as ants may be attracted to the honeydew it excretes. May be mistaken for aphids or whitefly, from which it differs by its large, mostly clear wings, and broad, cicada-like head and body.

Kristi Ellingsen

Kristi Ellingsen

LEFT Adult; RIGHT Nymph

ALYDIDAE (BROAD-HEADED BUGS)

Rice Seed Bug ■ *Leptocorisa acuta* TL 17mm
(Paddy Bug)

Rachel Whitlock

DESCRIPTION Elongated, slender, light green to yellow-brown body, with long, thin legs and antennae. Adults winged. **DISTRIBUTION** North-eastern Australia, and Asia-Pacific region and Central America. **HABITAT AND HABITS** Commonly found near watercourses or moist areas on grasses. Feeds in groups on developing seeds. If disturbed capable of emitting a foul odour. Adults can fly, but are not strong fliers. Pest of grain and grass crops, especially rice, and known to spread plant diseases.

Belostomatidae (Giant Water Bugs and Toe Biters)

Giant Water Bug ■ *Lethocerus (Lethocerus) insulanus* TL 50–70m
(Toebiter, Giant Fish-killer)

DESCRIPTION Large, flattened, oval-shaped insect. Dark brown mottled with lighter brown markings. Relatively small head with large, dark eyes, large raptorial forelimbs on either side of head, and other legs further back along body. Adults winged. **DISTRIBUTION** North-eastern Australia, from NT to NSW. Also Melanesia and the Philippines. **HABITAT AND HABITS** Found in fresh water, preferring calmer sites such as lakes and dams. Siphon at end of abdomen used like a snorkel to breathe. Ambush predator, feeding on a variety of aquatic prey such as arthropods, crustaceans, tadpoles and small fish. Uses piercing mouthparts to subdue prey and inject digestive fluids, and capable of giving a painful bite if handled. Adults can fly, and are often attracted to artificial lights. Insects in this family are the largest in the order Hemiptera.

Ryan Francis

Cimicidae (Bed Bugs or Cimicids)

Common Bed Bug ■ *Cimex lectularius* TL 5mm

DESCRIPTION Orange-brown to reddish-brown, with an oval-shaped, wingless body. When not feeding, proboscis is held flat against underside of thorax. **DISTRIBUTION** Found in close association with dense populations of humans, particularly in capital cities, and readily transported between locations. **HABITAT AND HABITS** Occurs in small groups in sleeping areas of host. Spends most of the day and night sheltering in a dark crevice, emerging for a few hours before dawn to feed on host's blood. Female needs blood for egg production. Eggs laid in small batches in crevices near sleeping area of host. Not responsible for major disease transmission, but severe bed bug infestations can lead to iron deficiency in people.

Piotr Naskrecki (cc)

Coreidae (Tip Wilters, Leaf-footed and Squash Bugs)

Donald Hobern (cc)

Crusader Bug

■ *Mictis profana* TL 25mm
(Holy Cross Bug)

DESCRIPTION Body dark brown to grey, with pale antennae tips and markings on abdomen. Adults have wings and a pale cross on back. Hind pair of legs thicker than first two pairs. **DISTRIBUTION** Across Australia. Also Indonesia and Indo-Pacific region. **HABITAT AND HABITS** Occurs in a range of habitats, including open woodland, heathland, parks and gardens. Feeds on sap of various plants, from native eucalypts and wattles, to cultivated citrus, grapes and roses. Used as biological control to combat the Giant Sensitive Tree *Mimosa pigra*, an invasive weed.

Corixidae (Water Boatmen)

Water boatmen ■ Multiple species TL 10mm

DESCRIPTION Streamlined, oval-shaped body with broad head and large eyes. Colour typically various shades of brown. Rear 2 pairs of legs longer than front ones, and paddle-like. Adults winged. **DISTRIBUTION** Worldwide. **HABITAT AND HABITS** Occurs in a wide range of freshwater habitats, favouring calmer waters such as lakes. Feeds mostly on algae and plant material, though a few species are predatory. Short front legs are used for feeding, while long rear legs are used for swimming. Adults are good fliers.

Larney Grenfell

Kristi Ellingsen

LEFT Dorsal view; RIGHT Swimming upside-down in water (ventral view)

GERRIDAE (WATER STRIDERS AND POND SKATERS)

Water Strider ■ *Limnogonus (Limnogonus) luctuosus* TL 11mm

DESCRIPTION Dark brown body with pale stripes, pale legs, and rear 2 pairs of legs very long and slender. Adults have winged and wingless forms. **DISTRIBUTION** North-eastern Australia. Also PNG, Melanesia and Polynesia. **HABITAT AND HABITS** Occurs in fresh water, from lakes and ponds, to streams and hot springs. Modified rear legs used to walk on water using surface tension. Predator, feeding on a range of small arthropods.

Peter Rowland/kapeimages.com.au

LYGAEIDAE (SEED BUGS)

Seed-eating bugs ■ *Graptostethus* spp. TL 10mm

DESCRIPTION Entire insect covered in bold black and orange-red patches and stripes. Adults winged; nymphs brightly coloured with large abdomen. **DISTRIBUTION** Throughout Australia, typically in warmer regions. Also Asia, Africa and Europe. **HABITAT AND HABITS** Commonly found in large groups feeding on seed pods of milkweed and other plants with toxic sap. Also feeds on seeds of various other plants, including garden plants and crops. Its aposematic colours warn predators that it is poisonous, sequestering toxins from the plant sap it feeds on. Many species of this family are involved in mimicry complexes with other bug species, and can be difficult to identify.

Bernhard Jacobi

Notonectidae (Backswimmers)

Backswimmers ■ Multiple species TL 5-20mm

DESCRIPTION Streamlined oval or boat-shaped, yellow-orange to brown body often with light and dark markings, with relatively small, broad head and large eyes. Front 2 pairs of legs short; hind pair much longer and fringed with hairs. Adults can be winged. **DISTRIBUTION** Throughout Australia. Worldwide. **HABITAT AND HABITS** Occurs in calm fresh water. Often characteristically seen 'hanging' upside-down below the water's surface as it exchanges air; can also swim upside-down. Adults can fly, and are capable of dispersing to temporary pools and ponds. Predatory and feeds on small invertebrates, can also feed on small fish and tadpoles.

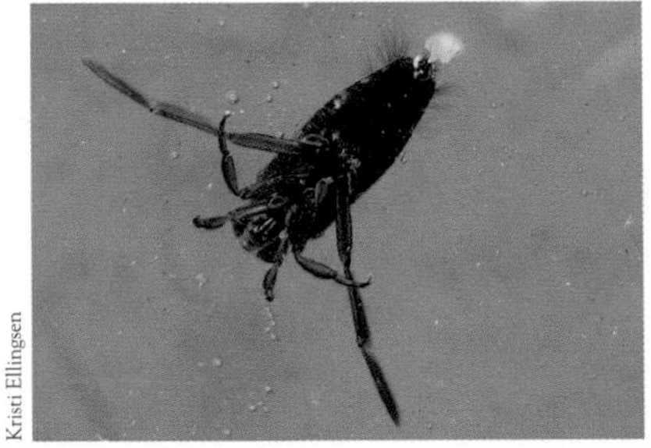

Kristi Ellingsen

Kristi Ellingsen

LEFT Swimming upside-down in water (ventral view); RIGHT Dorsal view.

Pentatomidae (Shield and Stink Bugs)

Spined Citrus Bug ■ *Biprorulus bibax* TL 20mm

Donald Hobern (cc)

DESCRIPTION Green, trapezoidal, shield-shaped body. Sometimes yellowish or pinkish, with black spines on 'shoulders' of thorax and small head. Nymphs rounder, wingless, with more extensive black markings. **DISTRIBUTION** Eastern Australia. **HABITAT AND HABITS** Native to inland regions of eastern Australia, where nymphal stages and adults feed on juices of young plant shoots and fruits in all stages of development. Host plants are Desert Lime *Eremocitrus glauca* and other native citruses. With the spread of commercial citrus orchards and these trees' popularity in domestic gardens, it has capitalized on the increased availability of food and become a pest in many areas. Once hatched, first-stage nymphs initially feed together, but soon disperse.

Green Vegetable Bug ■ *Nezara viridula* TL 15mm

DESCRIPTION Nymphs round, and dark brown to black, with pale white and orange markings. Later instar nymphs and adults typically green to brownish, sometimes orange. Adults more shield shaped, have a broad scutellum with 3 small white spots and are winged. **DISTRIBUTION** Worldwide. Thought to originate in Ethiopia. **HABITAT AND HABITS** Typically associated with agriculture and human activity. Feeds on a wide variety of host plants, but prefers legumes such as beans, including soybeans. Particularly attracted to plants that are fruiting or forming seed pods, and can form large populations that cause damage to crops, so considered an agricultural pest. Adults are strong fliers.

Simon Grove

Kristi Ellingsen

LEFT: Adult; RIGHT: Late instar nymph

Zebra shield bugs

■ *Poecilometis* spp. TL 15–25mm
(Gum tree shield bugs)

DESCRIPTION Trapezoidal, shield-shaped body, dark brown to rusty-red in colour, with black-and-white stripes or spotting, and dark wings covering tip of abdomen. Antennae yellow-orange and black, often banded. Nymphs typically rounder, wingless, and dark brown-black with pale stripes and other markings. **DISTRIBUTION** Throughout Australia. **HABITAT AND HABITS** Occurs in a variety of wooded habitats such as sclerophyll forests. Feeds on sap of native trees, preferring gum trees, banksia and native pines. Often found on leaves, trunks or under bark. Capable of producing a foul-smelling chemical when disturbed.

Rachel Whitlock

Gum Tree Shield Bug ■ *Theseus modestus* TL 15mm

DESCRIPTION Trapezoidal, shield-shaped body, dark brown with fine pale stripes and spotting; large yellow spot between wings on tip of scutellum. Antennae and legs black with yellow bands; dark wings. Head has 5 pale longitudinal stripes. Nymphs rounder, wingless. **DISTRIBUTION** Throughout Australia. **HABITAT AND HABITS** Occurs in open woodland areas including parks, feeding on sap of mostly eucalypts. Often spotted on leaves or trunks. Nymphs often found under bark. Capable of producing a foul-smelling chemical when disturbed. May be distinguished from similar *Poecilometis* spp. (see p. 69) by small size, and short wings that do not cover tip of abdomen and cerci.

Graham Wise (cc)

Reduviidae (Assassin Bugs)

Common Bee Killer Assassin Bug ■ *Pristhesancus plagipennis* TL 25mm

DESCRIPTION Yellowish to orange-brown and dark grey as adult, with transparent wings, long head, large eyes and thickened curved rostrum (proboscis). **DISTRIBUTION** Mainly coastal eastern Australia, from northern Qld, to around Sydney NSW. **HABITAT AND HABITS** Occurs in open forests, woodland and similarly vegetated areas, including urban parks and gardens, and cropland. Hunts among foliage for soft-bodied arthropods, particularly the European Honeybee (see p. 142) and caterpillars, which it grasps with its front legs and stabs with its powerful rostrum, injecting venom that paralyses and liquefies its prey before sucking out the body fluids. Capable of inflicting an intense painful bite.

Peter Rowland/kapeimages.com.au

SCUTELLERIDAE (JEWEL AND METALLIC SHIELD BUGS)

Ground Shield Bug ■ *Choerocoris paganus* TL 12mm
(Red Jewel Bug)

DESCRIPTION Convex, shield-like body, bright red to orange and yellow, with metallic blue-green head and legs, and metallic blue-green patches on thorax and abdomen. **DISTRIBUTION** Throughout Australia. **HABITAT AND HABITS** Feeds on sap, particularly from seeds and leaves, of various native plants. Often found in large congregations on rocks or on the ground, feeding on fallen seeds and small tussocks of grass. May also feed on introduced and cultivated plants such as alfalfa. May be mistaken for a beetle due to its bright metallic colouration, but lacks hard wing cases (elytra) of beetles, and has typical sucking mouthparts of true bugs.

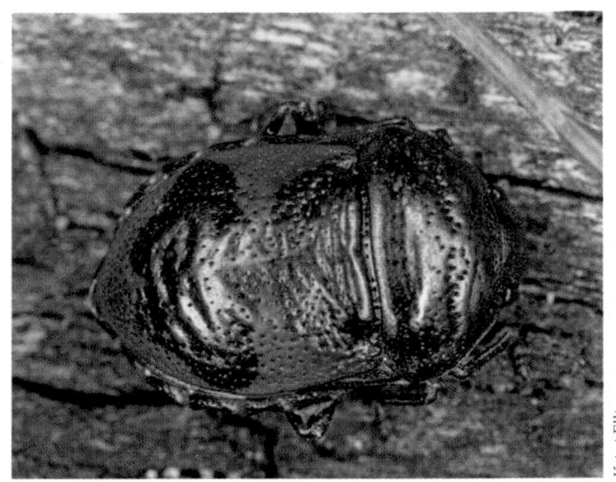
Kristi Ellingsen

Metallic Shield Bug ■ *Scutiphora pedicellata* TL 15mm

DESCRIPTION Convex, shield-like body, bright metallic green with darker green-blue markings, and bright yellow to orange-red markings on scutellum and abdomen. **DISTRIBUTION** Coastal eastern Australia. **HABITAT AND HABITS** Usually occurs in forested areas under bark and on the ground, though seems to prefer cultivated plants and often found on vegetables, fruit trees and ornamental garden plants, feeding on sap from leaves, stems, fruits and grains. Gregarious and can be found in groups of all life stages, with particularly large congregations occurring at breeding times. May be mistaken for a beetle due to its bright metallic colouration, but lacks rigid wing cases (elytra) of beetles, and has typical sucking mouthparts of true bugs.

Peter Rowland/kapeimages.com.au

Hibiscus Harlequin Bug ■ *Tectocoris diophthalmus* TL 20mm
(Cotton Harlequin Bug)

Peter Street

DESCRIPTION Convex, shield-like body. Adults bright orange-yellow to red with variable metallic blue-green markings; old individuals duller in colour. Nymphs round, wingless, with more extensive blue-green colour. **DISTRIBUTION** Northern and eastern Australia. Also PNG and Asia-Pacific region. **HABITAT AND HABITS** Occurs in a range of habitats, from agricultural areas to coastal forests. Feeds mainly on plants in hibiscus family, Malvaceae, including cotton, as well as native *Brachychiton* spp., grevillea and bottlebrush. Most commonly seen in summer when congregations of insects feed on sugar-rich sap of new shoots, buds and seeds. Large populations may become minor pests in gardens or cotton plantations, causing leaves and buds to drop, or by spreading plant pathogens. May be mistaken for a beetle due to its bright metallic colouration, but lacks rigid wing cases (elytra) of beetles, and has typical sucking mouthparts of true bugs.

TINGIDAE (LACE BUGS)

Lantana Lace Bug ■ *Teleonemia scrupulosa* TL 4mm

DESCRIPTION Tiny, elongated bugs, mottled brown in colour, with humped thorax, 'X' shape formed by edges of scutellum, and tegmen of wings visible from above. Nymphs wingless, with spiny head and abdomen. **DISTRIBUTION** Introduced to Australia. Native to Central America, and introduced to Africa, Asia and Oceania. **HABITAT AND HABITS** Brought to Australia and many other areas as a biological agent to control populations of the invasive weed *Lantana*. Usually feeds on undersides of leaves and does not tend to affect other plant species. Black droplets of excreta, old shed nymphal exoskeletons and pale dying leaves may indicate its presence. Populations become most abundant seasonally in warm, dry areas.

Steve and Alison Pearson

ALEYRODIDAE (WHITEFLIES)

Silverleaf Whitefly ■ *Bemisia tabaci* TL 1–2mm

(Cotton Whitefly, Sweetpotato Whitefly, Tobacco Whitefly)

DESCRIPTION Adults are tiny flying insects, with a pale yellowish body and 4 white waxy wings. Round, flat, translucent nymphs resemble scale insects. **DISTRIBUTION** Worldwide, in tropical, subtropical and warm temperate regions. **HABITAT AND HABITS** Typically found in areas of agriculture, in greenhouses and gardens. Polyphagous, sucking sap of a variety of plants. Heavy infestations can damage host plants through feeding activity and sooty mould caused by excreted honeydew. Also a vector for plant pathogens, making it an important agricultural pest.

CSIRO Scienceimage (cc)

Greenhouse Whitefly ■ *Trialeurodes vaporariorum* TL 2mm

(Glasshouse Whitefly)

DESCRIPTION Adults are tiny flying insects, with a pale yellowish body and white waxy wings. Round, flat nymphs resemble scale insects and vary in colour from green to dark brown. **DISTRIBUTION** Worldwide. **HABITAT AND HABITS** Typically found in areas of agriculture, in greenhouses and gardens. Polyphagous, sucking sap of a variety of plants. Heavy infestations can damage host plants through feeding activity and sooty mould caused by excreted honeydew. Also a vector for plant pathogens, making it an important agricultural pest.

Guido Bohne (cc)

APHIDIDAE (APHIDS)

Oleander Aphid ■ *Aphis (Aphis) nerii* TL 3mm
(Milkweed Aphid)

DESCRIPTION Rounded, bright orange-yellow body; legs and long antennae black. Small head and large abdomen, with 2 long black tubes on rear of abdomen. Winged and wingless forms occur. **DISTRIBUTION** Introduced to Australia. Worldwide, in tropical, subtropical and warm temperate regions. **HABITAT AND HABITS** Feeds primarily on sap of plants in the Apocynaceae family, such as oleanders and milkweeds, and occasionally also on plants in other families. Typically infests ornamental plants, though can be a vector for many plant viruses that affect agricultural crops.

Dianne Clarke

Rose Aphid ■ *Macrosiphum (Macrosiphum) rosae* TL 4mm

DESCRIPTION Rounded body ranging in colour from green to pink and light brown. Legs and long antennae pale with black patches. Younger nymphs lighter in colour. Small head and large abdomen, with 2 long, dark tubes on rear of abdomen. Winged and wingless forms occur. **DISTRIBUTION** Almost worldwide; largely absent from eastern Asia. **HABITAT AND HABITS** Typically feeds on sap of roses *Rosa* spp. in large colonies, favouring soft growing tips and buds. Can infest other species in the Rosaceae family, and occasionally plants in other families. Populations can grow rapidly, and large infestations can cause damage to host plants.

Bernard DuPont (cc)

Coccidae (Soft and Wax Scales)

White Wax Scale ■ *Ceroplastes destructor* TL 6mm
(Soft Wax Scale, Citrus Waxy Scale)

DESCRIPTION Rounded, with marginal setae, small legs, and body covered in thick layer of pale white, smooth wax that is soft to the touch. **DISTRIBUTION** Introduced to Australia. Worldwide. **HABITAT AND HABITS** Feeds on various plant hosts, including citruses and other fruit crops, and considered an agricultural pest. Large infestations can damage plants and affect crop yields. May also feed on native plants such as wattles and lilly-pilly. Adult females typically immobile.

Plant & Food Research

Wattle Tick Scale ■ *Cryptes baccatus* TL 10mm

DESCRIPTION Round, smooth, sack-like body covered in wax, light blue-grey to grey-pink in young insects, changing to brown in older insects. **DISTRIBUTION** Throughout Australia. **HABITAT AND HABITS** Forms large congregations on branches and stems of wattles, where it feeds on sap. Ants are often found tending it and feeding on the excreted honeydew. Adult insects immobile.

Peter Clark

DACTYLOPIIDAE (COCHINEALS)

Prickly Pear Cochineal ■ *Dactylopius opuntiae* TL 3mm
(Opuntia Cochineal Scale)

DESCRIPTION Larger females wingless, with dark reddish, rounded or oval body, typically covered in white sticky wax. Smaller males winged. **DISTRIBUTION** Introduced to Australia. Tropical and subtropical regions worldwide, originating in Central America. **HABITAT AND HABITS** Feeds on prickly pear cacti *Opuntia* spp., and was introduced to Australia as a biocontrol agent for these invasive cacti. In other countries it is a pest of cultivated species of prickly pear. It is in the same genus as the insect used for red cochineal dye, *D. coccus*, and is also a bright red colour when crushed.

Vahe Martirosyan (cc)

ERIOCOCCIDAE (FELT SCALES AND ERIOCOCCIDS)

Eucalyptus galling scales ■ *Apiomorpha* spp. TL 2–45mm
(Eucalyptus felt scales, Gumtree galls)

DESCRIPTION Adult females pale, soft, teardrop-shaped, with very short legs, usually covered in white powdery wax. Adult males much smaller, winged, with long filaments at tip of abdomen. **DISTRIBUTION** Throughout Australia. Also PNG. **HABITAT AND HABITS** Occurs on numerous species of gum tree, on which it induces galls grown from the plant tissue. Feeds on sap and plant tissue inside gall. Adult females live in large galls on branches that vary greatly in shape and size, depending on the species, and may be mistaken for fruits of host tree. Males typically make small, narrow cylindrical galls on leaves of host. Species in this genus form some of the world's longest insect galls

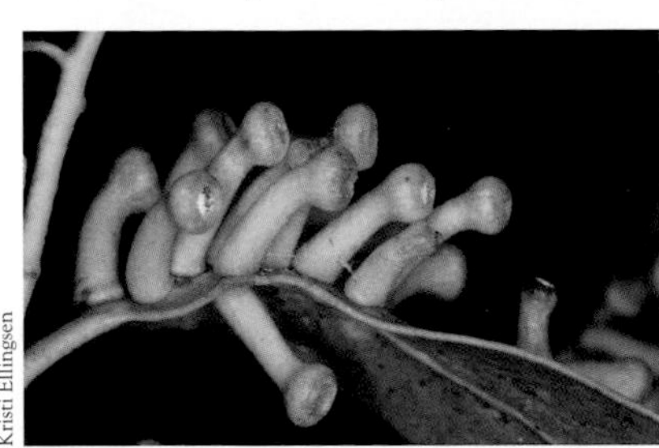

Kristi Ellingsen

Rachel Whitlock

Galls can vary greatly in shape and size.

MONOPHLEBIDAE (GIANT SCALES AND MONOPHLEBIDS)

Giant snowball mealybugs ■ *Monophlebulus* spp. TL 10–25mm

DESCRIPTION Body of nymphs and adult females oblong, soft, wingless, and blue-grey in colour with bright orange markings. Produces white wax over body with fine white, hair-like filaments. Adult males smaller, purplish in colour, with long, dark antennae and dark purple wings. **DISTRIBUTION** Eastern and southern Australia. Also Southeast Asia **HABITAT AND HABITS** Typically found on *Eucalyptus* and *Callistemon* branches or leaves, where it feeds on plant sap. Able to move slowly around host plant, and ants may feed on honeydew it secretes. Larger, older females may have less of the white waxy covering than younger insects.

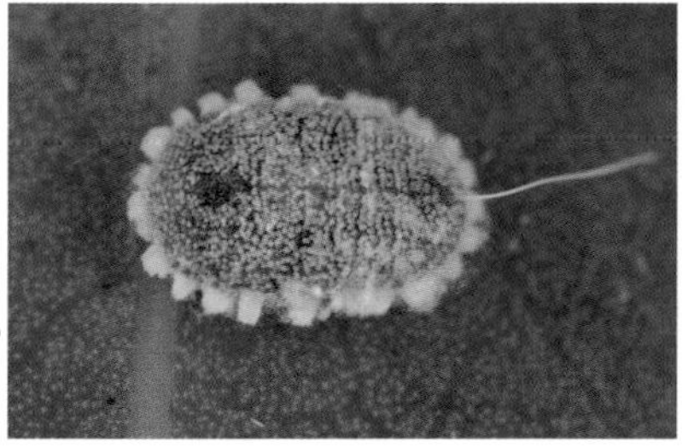

Kristi Ellingsen

Simon Grove

LEFT Nymph; RIGHT Adult.

PSEUDOCOCCIDAE (MEALYBUGS)

Long-tailed Mealybug

■ *Pseudococcus longispinus* TL 5mm

DESCRIPTION Soft pinkish, wingless body with segmented appearance; produces white powdery wax covering body. White filaments around margin of body, with distinctive long abdominal filaments in adult females. Adult males dark, slender and winged. **DISTRIBUTION** Worldwide, in tropical, subtropical and warm temperate regions. **HABITAT AND HABITS** Cosmopolitan pest that is polyphagous, feeding on sap of a wide variety of ornamental and agricultural plants. Honeydew it secretes can attract attendant ants, and can cause sooty mould to grow on the host plant. Can also transmit diseases between plants. Heavy infestations appear as waxy white masses on affected plants, and can cause damage to crops such as citruses, pears and macadamias.

D. Kuru (cc)

PSYLLIDAE (JUMPING PLANT LICE OR PSYLLIDS)

Lace lerps ■ *Cardiaspina* spp. TL 4mm
(Basket lerps)

DESCRIPTION Very small insect, yellow-orange to dark brown in colour, often striped. Adults winged, resembling tiny cicadas. Lerps of this genus are typically 1–4mm wide, pale white to brown in colour, and resemble lace, or a basket or shell. **DISTRIBUTION** Throughout Australia, including Norfolk Island. Introduced to NZ. **HABITAT AND HABITS** Feeds primarily on sap of gum trees, and commonly found on undersides of leaves. Heavy infestations can cause significant defoliation of trees. Nymphs create a protective cover out of sugars and amino acids obtained from plant sap. The case is called a lerp, although 'lerp' often also refers to the insect, and its colour, size and shape differs among species.

Rachel Whitlock

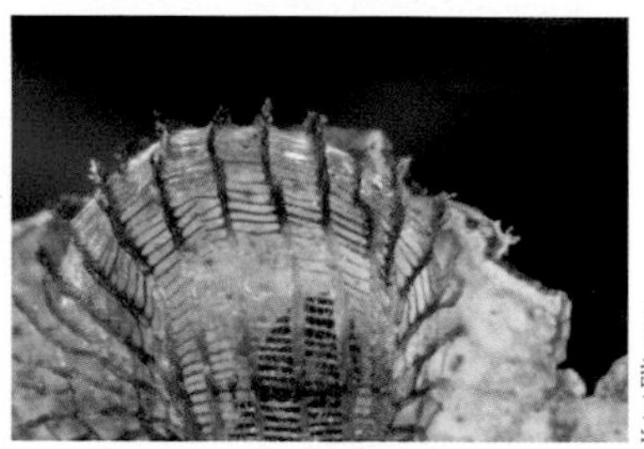

Kristi Ellingsen

LEFT Underside of leaf showing lerp infestation; RIGHT Close-up reveals concealed insect.

Lou Wolfers

ASCALAPHIDAE (OWLFLIES)

Common Owlfly
■ *Suhpalacsa subtrahens* TL 45mm

DESCRIPTION Adults have yellow and grey-black body, long antennae with clubbed tips and transparent wings. Female has thicker abdomen than male. Larvae have rounded hairy body, and large mandibles with multiple teeth. **DISTRIBUTION** Coastal eastern and northern Australia. **HABITAT AND HABITS** Usually spotted at rest on plant stems, with wings held downwards and abdomen held out at an angle to mimic a twig. Larvae forage in leaf litter or on trees. Both larvae and adults are predatory. Commonly confused with dragonflies or damselflies, it can be easily distinguished by the position of its wings at rest and its long antennae.

Chrysopidae (Green Lacewings)

Green lacewings ■ Multiple species TL 20mm
(Common lacewings)

DESCRIPTION Adults have light brown to yellow and bright green bodies, mostly transparent, net-like wings and long, thin antennae. Larvae typically brownish, though can vary in colour, hairy, with obvious mandibles. **DISTRIBUTION** Worldwide. **HABITAT AND HABITS** Occurs in a range of environments, typically around vegetation but wherever prey arthropods can be found. Larvae free moving, and usually cover themselves with debris in order to ambush prey. Larvae and adults predate mostly on small, soft-bodied insects such as aphids and mealybugs, and are widely used as a biological control of many common garden pests. Flying adults may be attracted to lights at night. There are many species in the family; most look similar and require specialist identification to tell apart.

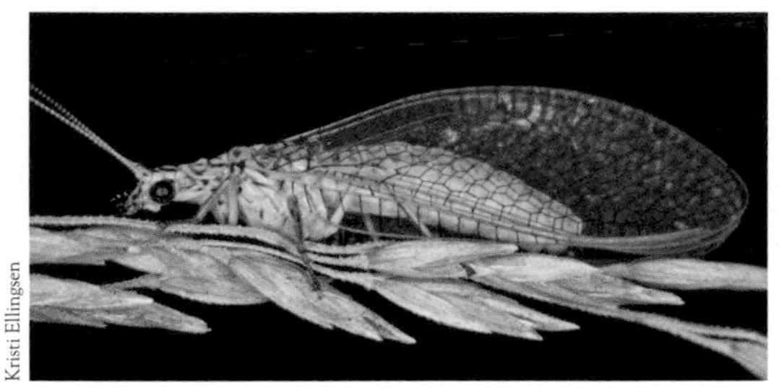

Kristi Ellingsen

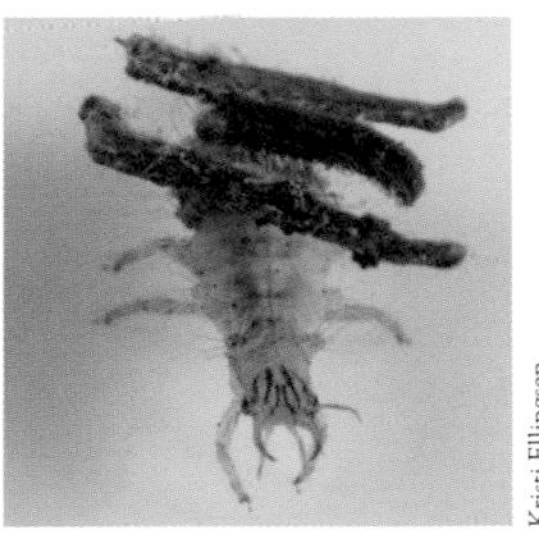

Kristi Ellingsen

LEFT Adult; RIGHT Larvae carrying debris for camouflage

Mantispidae (Mantid Lacewings and Mantisflies)

Wasp-mimicking mantisflies ■ *Austromantispa* spp. TL 30mm

DESCRIPTION Adults vary in colour from mottled brown-grey to black and orange-yellow. Adults have mostly transparent wings, short antennae and distinctive raptorial forelimbs, similar to those of preying mantises. **DISTRIBUTION** Throughout Australia. **HABITAT AND HABITS** Typically found on plants and in leaf litter. Larvae predatory, feeding on egg sacs and young of other arthropods. Adults also predatory and utilize their raptorial forelimbs to capture prey.

Ryan Francis

MYRMELEONTIDAE (ANTLIONS)

Antlion lacewings ■ *Myrmeleon* spp. TL 60mm

DESCRIPTION Adults have long, slender body, dark grey-brown in colour, short antennae with clubbed or hooked tips, and mostly transparent, narrow wings. Larvae rounded, mottled brown, hunchbacked in appearance, with obvious mandibles with multiple teeth. **DISTRIBUTION** Throughout Australia. **HABITAT AND HABITS** Antlion larvae construct typical sand traps up to 40mm in diameter in loose soil. When an ant or other small arthropod walks into the trap, the larva flicks sand until it causes a small landslide, sending the ant to the bottom of the trap, where it is seized by the larva's large mandibles; then digestive enzymes are injected into it. Adults also predate on small arthropods, and are nocturnal; they are sometimes attracted to lights at night.

Kristi Ellingsen

Kristi Ellingsen

LEFT: Exposed larvae; RIGHT: Larvae concealed within sand trap

Hairy antlion lacewings ■ *Heoclisis* spp. TL 50mm

Donald Hobern (cc)

DESCRIPTION Adults range from dark to pale grey, with black or yellow markings, large, dark eyes, short antennae with clubbed tips, mottled wings, and body with a hairy or shaggy appearance. Typical antlion larvae. **DISTRIBUTION** Throughout Australia. **HABITAT AND HABITS** Prefers dry areas with loose soil, in which larvae construct typical antlion-type sand traps to 40mm in diameter. Larvae and adults both predatory, feeding on small arthropods.

Nymphidae (Split-footed Lacewings)

Giant Orange Lacewing ■ *Nymphes myrmeleonoides* TL 30mm; WS 90mm

(Blue Eyes Lacewing)

DESCRIPTION Adults have yellow-orange body, large, blue-black eyes, black dorsal stripe on abdomen, long, thin antennae that are black with orange tips, and long, narrow wings that are transparent with white tips. Larvae large, rounded, antlion-like; long, thin mandibles with single tooth halfway along; round, hairy body typically covered with debris particles. **DISTRIBUTION** Eastern Australia. **HABITAT AND HABITS** Inhabits open woodland. Adults and larvae both predatory, with larvae ambushing prey among leaf litter, and adults capable of ambushing prey on the wing. Adults may fly into houses, attracted by lights at night. Females lay white eggs on ends of long, thin threads in a distinctive 'U' shape, often beneath overhangs around buildings. Capable of emitting a pungent odour when disturbed.

John Tann (cc)

Corydalidae (Dobsonflies)

Dobsonflies ■ *Archichauliodes* spp. TL <50mm; WS <80mm

DESCRIPTION Adults usually brown in colour, with spotted wings, and may have orange bands on thorax. Larvae large (<30mm) with dark head, obvious mandibles, and 8 pairs of gills on abdomen that may be similar in appearance to legs. **DISTRIBUTION** Eastern Australia and south-west WA. Also NZ. **HABITAT AND HABITS** Larvae aquatic, and usually found under rocks and among mud and debris in cold freshwater systems. Larvae are ambush predators of other invertebrates, while adults are short-lived and probably do not feed. Larvae hatch from eggs laid on vegetation close to water and go through up to 12 moults before reaching adult stage and leaving the water.

Shaun Winterton

Sialidae (Alderflies)

Alderfly ■ *Stenosialis australiensis* TL <25mm
(Toebiter)

DESCRIPTION Adults have orange-brown bodies, and brown wings with dark spotting. Larvae have 7 pairs of gills on abdomen that may be similar in appearance to legs, and long filament at end of abdomen. **DISTRIBUTION** Through coastal and near-inland eastern Australia from northern Qld, to central Vic. **HABITAT AND HABITS** Typically occurs in clear, cold waters, but may also be found among silt and mud. Adult female lays large numbers of eggs on grass stems near water. Larvae usually found under rocks and debris, adults on stems of riparian vegetation, and are active predators of other invertebrates. Larvae take up to 2 years to reach adult stage, but adults live for only 2–3 weeks.

Shaun Winterton

Carabidae (Tiger Beetles and Carabids)

Green Carabid ■ *Calosoma (Australodrepa) schayeri* TL 20mm
(Saffron Beetle)

DESCRIPTION Metallic green beetle with black legs and head, and long antennae. Elytra rectangular in shape with pitting in longitudinal rows. Broad pronotum narrows where it meets head and elytra, giving waist-like appearance similar to a hot-water bottle. Larvae have an elongated, shiny black, soft body, pale head and developed legs. **DISTRIBUTION** Throughout Australia, but largely absent from northern wet tropics. **HABITAT AND HABITS** Widespread in most habitat types except tropical northern reaches of Qld, NT and WA. Feeds on slow-moving invertebrates, largely caterpillars, actively hunting them at night. Both adult beetle and larva are predatory on the ground in leaf litter; adults also able to climb and fly. May be attracted to lights at night. Capable of producing a foul smell if disturbed.

Dianne Clarke

BUPRESTIDAE (JEWEL BEETLES)

Jewel beetles ■ *Castiarina* spp. TL 10–20mm

DESCRIPTION Elongated beetle; elytra have longitudinal grooves, tapering to tip of abdomen. Head smaller than broad pronotum, with large eyes. Occurs in a wide variety of vivid metallic colours, including black, yellow, red, blue and iridescent green-purple, often with blotchy markings on elytra that vary in and between species. Larva is a white grub with flattened head. **DISTRIBUTION** Throughout Australia. **HABITAT AND HABITS** Diverse and colourful genus of jewel beetles, with more than 400 species occurring across Australia in a variety of habitats. Some species appear quite similar and require specialist identification to differentiate. Adult beetles generally emerge in summer, coinciding with flowering season of native plants such as gum trees and tea trees *Leptospermum* spp. Larvae eat plant matter, usually boring into wood, roots or plant stems.

Kathryn Himbeck

Variable Jewel Beetle ■ *Themognatha variabilis* TL 35mm

DESCRIPTION Large, elongated beetle, tapering to tip of abdomen; body green to black with elytra varying from bright yellow to dark red-brown; markings on either side of pronotum match elytra colour. Elytra ridged longitudinally, sometimes with black or green markings. **DISTRIBUTION** South-east Australia. **HABITAT AND HABITS** Found in a range of forested habitats where its food trees are available. Feeds on flowers of native plants mostly in the Myrtaceae family, such as gum trees and tea trees *Leptospermum* spp. Larvae are wood borers, and have been found in she-oak trees *Casuarina* spp. Adults usually only active for brief period during warmer months each year.

John Tann (cc)

Cantharidae (Soldier and Leatherwing Beetles)

Soldier beetles ■ *Chauliognathus* spp. TL 10–20mm

DESCRIPTION Elongated beetle with soft body, and leathery rectangular elytra often only covering part of abdomen. Many species in the genus, which vary in their contrasting colours, including black and yellow, blue and red, and orange and metallic green. **DISTRIBUTION** Throughout Australia, including Lord Howe Island. Also PNG, and North and South America. **HABITAT AND HABITS** Found in a range of woodland habitats, including dry sclerophyll forests, parks and gardens. Adult beetles feed on flower nectar and pollen. Larvae live among leaf litter or in soil and feed on other invertebrates. Bright colours of beetles serve as warning to predators, as they are capable of producing a milky white defensive chemical if threatened. Beetles of some species, such as the **Plague Soldier Beetle** *C. lugubris* (pictured), may congregate in very large numbers, covering entire trees.

Andrew Allen

Cerambycidae (Longicorn or Longhorn Beetles)

Poinciana Longicorn ■ *Agrianome spinicollis* BL 60mm

DESCRIPTION Large beetle; oblong-shaped body reddish-brown; elytra light brown and partially translucent. Sizeable mandibles; pronotum edged with small spines and long antennae. Larvae large white grubs with dark head and no obvious legs. **DISTRIBUTION** Eastern Australia and Lord Howe Island. **HABITAT AND HABITS** Occurs in rainforests, sclerophyll forests, parks and other wooded areas. Larvae are wood boring, usually in dead and rotting wood of poinciana, fig and pecan trees. May become a pest of these, especially pecan trees. Large, formidable-looking adult often encountered as it blunders into houses, attracted to lights at night. Its folded hindwings may be partially visible under its relatively thin elytra. Larvae are one of a few species of insect larvae sometimes eaten as witchetty grubs.

Scott Eipper

Tiger Longicorn ■ *Aridaeus thoracicus* BL 20mm

DESCRIPTION Elongated beetle; contrasting black-and-orange markings on body; femur of each leg swollen. Antennae same length as body, longer in males than females. Larvae pale grubs with dark head. **DISTRIBUTION** Eastern Australia. Introduced to NZ. **HABITAT AND HABITS** Found in woodland habitats, parks and gardens, usually during warmer months. Adult beetle feeds on flower nectar and pollen, typically of native plants such as gum trees and bottlebrushes, though also feeds on flowers of introduced plants including citrus and pear. Larvae bore into various species of dead trees. Colouration and behaviour of adult suggests it may be a wasp mimic, and it is often mistaken for a wasp.

John Tann (cc)

Common Eucalypt Longicorn ■ *Phoracantha semipunctata* BL 15–25mm

(Eucalyptus Longhorn Borer)

DESCRIPTION Elongated dark brown to reddish-brown beetle with yellowish patches in middle and tips of elytra; elytra wider than head and pronotum, long legs and sizeable jaws. Antennae longer than body length, longest and spined in males. Larvae pale grubs with reddish-brown head and no obvious legs. **DISTRIBUTION** Throughout Australia. Introduced worldwide. **HABITAT AND HABITS** Native to Australia, occurring mainly in eucalypt woodland where host trees are abundant. Accidentally introduced overseas, where it is a pest of eucalypt plantations. Larvae are wood borers and cause damage to trees, typically those already dying or under stress, resulting in death of branches or an entire tree in extreme cases. Adults can be found near or on eucalypts, usually under loose bark where eggs are laid. May be attracted to lights at night.

Andrew Allen

CHRYSOMELIDAE (LEAF AND TORTOISE BEETLES)

Tortoise beetles ■ *Aspidimorpha* spp. TL 10mm

DESCRIPTION Rounded, somewhat flattened beetle, with flat 'skirt' on elytra and pronotum obscuring head and legs. Occurs in various colours from cream to orange-yellow or green, with varying darker patterns including mottling and spotting, and skirt often translucent. Larvae oval, flattened, often with spiny projections around body margin; some carry past shed exoskeletons. **DISTRIBUTION** Northern and eastern Australia. Also Southeast Asia. **HABITAT AND HABITS** Found in a range of habitats, usually forested areas, where adults and larvae feed on leaves of shrubs and trees. When threatened, tucks legs and antennae underneath body and clamps down. Larvae that retain past exoskeletons can use these as decoys or weapons to fend off potential predators.

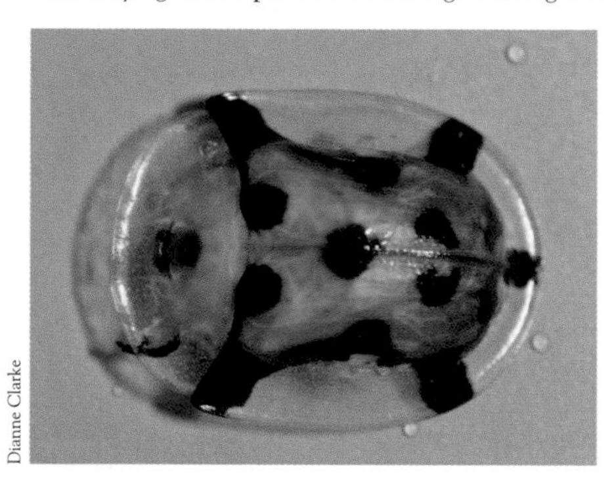

Dianne Clarke

Acacia leaf beetles ■ *Dicranosterna* spp. TL 10–20mm

DESCRIPTION Rounded, domed beetle, varying in colour, including pale green, honey-brown, orange-red and black; often with pitting on elytra. Larvae have soft, pale body, dark sclerotized head and legs, and enlarged globular abdomen. **DISTRIBUTION** Throughout Australia. Introduced to NZ. **HABITAT AND HABITS** Found in a variety of habitats, but typically in wooded areas where its food trees are available. Adults and larvae feed primarily on leaves and flowers of various species of *Acacia*. When threatened, tucks legs and antennae underneath its strong domed body.

Reiner Richter

Southern Eucalyptus Leaf Beetle ■ *Paropsisterna agricola* TL 15mm

Kristi Ellingsen

DESCRIPTION Domed beetle, varying in colour from red-orange to gold or blue-grey, sometimes with red 'skirt', dark markings on head and pronotum, and lace-like pattern on elytra. **DISTRIBUTION** South-eastern Australia. **HABITAT AND HABITS** Found in eucalypt forests, and feeds on leaves and flowers of gum trees. Larvae also feed on leaves and young shoots. When large populations build up, can become a pest of eucalypt plantations grown for timber.

Coccinellidae (Ladybirds or Ladybeetles)

Tasmanian Ladybird ■ *Cleobora mellyi* TL 5–8mm
(Southern Ladybird)

DESCRIPTION Adult's head, prothorax and weakly convex elytra yellowish-orange to red, with numerous large black blotches and zigzag patterns. Larvae elongated, becoming mostly blue-back to black with yellow tubercles and yellow head. All stages have 6 legs. **DISTRIBUTION** Tas and southern mainland Australia, including south-western WA, south-eastern SA, Vic and eastern NSW. Introduced to NZ. **HABITAT AND HABITS** Found in acacia and eucalypt forests and woodland, and cultivated parks and gardens. Adults and larvae feed on eggs and larvae of tortoise beetles (see opposite) and other leaf beetles, and psyllid species. Adult female lays clusters of yellow eggs on plant stems near infestations of these prey species, and wingless larvae progress through 4 stages before pupating into adult form.

Andrew Allen

Kristi Ellingsen

LEFT Adult; RIGHT Larva.

Transverse Ladybird ■ *Coccinella transversalis* TL 4–6mm

DESCRIPTION Adults red or orange and black. Elytra dome shaped with evenly distributed, black elongated patches. Larvae greyish-brown with darker markings; pale or dark soft spikes on edges of each body segment and 6 legs. Body tapers at both ends (fusiform).

Peter Rowland/kapeimages.com.au

DISTRIBUTION Throughout Australia. Australia's most common ladybird beetle. **HABITAT AND HABITS** Occurs in forests, woodland, heaths, parks and gardens. Adults and young feed on invertebrates, including significant pests of agricultural and garden plants such as aphids, mites and scale insects. Can also feed on pollen. Both adults and larvae active during the day and can feed on the same plants; may be in the company of other ladybird species.

Mealybug Destroyer
■ *Cryptolaemus montrouzieri* TL 4–5mm

DESCRIPTION Adults spotless, with mainly blackish-brown elytra and orange-tan head and posterior, and covered in short, white fur. Larvae large (to 14mm), white and with waxy appendages on body, resembling large mealybugs. **DISTRIBUTION** Native to eastern Australia, including Qld and NSW. Introduced to other Australian states and territories, and several countries around the world, as biological control. **HABITAT AND HABITS** Found in a variety of vegetated habitats, including commercial agricultural areas and cultivated parks and gardens. Adults and larvae feed on mealybugs and other soft scale insects, and female lays yellow eggs in mealybug egg clusters or adjacent prey infestations.

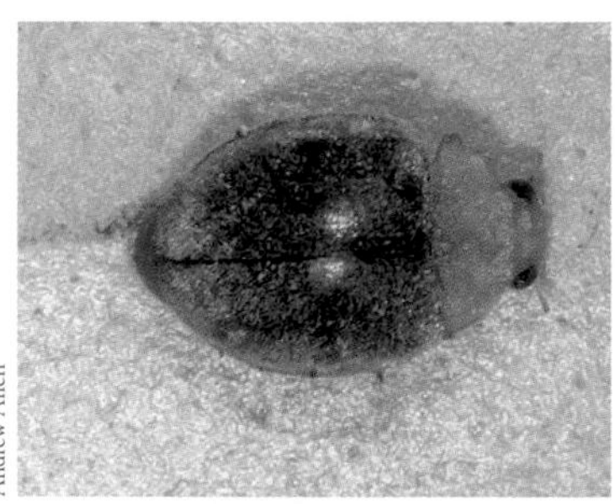

Andrew Allen

Andrew Allen

LEFT Adult; RIGHT Larva.

Steelblue Ladybird ■ *Halmus chalybeus* TL 3–4mm

DESCRIPTION Adult's head, prothorax and elytra shiny dark metallic bluish to greenish, with front of head and front corners of prothorax yellowish-brown in male. Head small with compound eyes and short antennae. Larvae elongated, and pale grey with long, dark fleshy extensions and long, hair-like seta at tip. **DISTRIBUTION** Widely distributed in eastern Australia, from northern Qld to Vic. Introduced into NZ and the Americas. **HABITAT AND HABITS** Found in woodland and shrubland, including parks and gardens. Adults feed on various small insects, including scale insects, psyllids and whiteflies, and also on honeydew, nectar and some plant exudates. Females lay yellow eggs, some with dark material on top. When fourth larval instar is fully grown, it attaches itself to a protected place on a plant to pupate.

Peter Rowland/kapeimages.com.au

Common Spotted Ladybird ■ *Harmonia conformis* TL 7–10mm

DESCRIPTION Adults bright orange to reddish with about 20 large black spots on elytra and pronotum, some of which may pool together. Larvae blackish and ant-like after hatching, later developing yellow tubercles and two yellow-orange bands on abdomen. **DISTRIBUTION** Native to eastern and southern Australia, including Qld, NSW, Vic, Tas, south-eastern SA and south-western WA. Introduced to NZ. **HABITAT AND HABITS** Occurs in a range of vegetation types, including urban parks and gardens. Adults and larvae are predators of soft-bodied invertebrates such as aphids, scale insects and mites, but also feed on honeydew. Larvae progress through 4 stages, shedding outer skin between each stage, and growing to c. 12mm in length before pupating into adult form, which last for about 2 months.

Peter Rowland/kapeimages.com.au

Fungus-eating Ladybird ■ *Illeis galbula* TL 4–5mm

DESCRIPTION Adults bright yellow with broad, jagged black bands across elytra and black midline. Larvae have 6 legs; first segment of elongated body yellow, remainder grey, each segment conspicuously marked with black dots. **DISTRIBUTION** Eastern Australia. Introduced into NZ. **HABITAT AND HABITS** Occurs on trees and shrubs in vegetated habitats, including parks and gardens, where it is active during the day and fast moving. If disturbed, adults employ a drop-and-fly-away method to escape. Adults and larvae feed on fungi of the genus *Oidium*, which cause powdery mildew, and other fungi; adults may also feed on aphids, black mould and pollen.

Peter Rowland/kapeimages.com.au

Peter Rowland/kapeimages.com.au

LEFT Adult; RIGHT Larva.

CURCULONIDAE (WEEVILS AND SNOUT BEETLES)

Diamond Weevil ■ *Chrysolopus spectabilis* 25mm

(Botany Bay Beetle)

DESCRIPTION Adults blackish with emerald-green patches and lines over upper body, and greenish wash on underparts. Snout characteristically long, with thickened antennae towards tips, and eyes moderately large. Larvae legless. **DISTRIBUTION** South-eastern Australia, including southern Qld, NSW, Vic and eastern SA. **HABITAT AND HABITS** Found in acacia forests, woodland and where host trees grow in parks and gardens. Feeds on acacias; adults on smaller branches and leaves, and larvae on roots, presumably of same tree species. When disturbed, adults drop to the ground and play dead. One of the first Australian insects scientifically described, with specimens collected by Sir Joseph Banks in 1770.

Peter Street

Wattle pigs

■ *Leptopius* spp. TL 10–20mm

DESCRIPTION Adult beetle light grey-brown to dark brown in colour; pronotum and elytra have bumpy protrusions arranged in longitudinal ridges. Snout characteristically elongated, relatively thick, with clubbed tip; elbowed antennae; dark eyes moderately large. Larvae legless. **DISTRIBUTION** Throughout Australia. **HABITAT AND HABITS** Occurs in a range of habitats, from sclerophyll forests to urban gardens. Feeds primarily on wattle species, though some species may be found on other native plants like gum trees and bottlebrushes. Adults slow moving, and when disturbed resort to dropping to the ground and playing dead.

Peter Rowland/kapeimages.com.au

Elephant Weevil ■ *Orthorhinus cylindrirostris* TL 10–20mm

DESCRIPTION Adult beetle mottled with various shades of brown and cream and thick, dark scales; pronotum and elytra have ridges and bumpy protrusions. Forelegs longer and thicker than other legs, longest in males; snout characteristically long; elbowed clubbed antennae. Larvae pale, legless, with brown head. **DISTRIBUTION** Eastern Australia. **HABITAT AND HABITS** Inhabits woodland areas, where it feeds on eucalypts and other native trees. Has adapted well to feeding on various introduced plant species. Has become a pest especially in the wine industry, where the larvae bore into cultivated grapevines.

Kristi Ellingsen

Dermestidae (Dermestid and Carpet Beetles)

Carpet Beetles ■ *Anthrenus* spp. TL 2–4mm

DESCRIPTION Small, rounded beetle, patchy pattern formed by scales on body, ranging in colour, including black, brown, white, yellow and orange-red. Larvae reddish-brown and covered with hairs. **DISTRIBUTION** Introduced to Australia. Worldwide. **HABITAT AND HABITS** Typically associated with human activity. Adults feed on pollen and nectar, then deposit eggs in a variety of sites including carrion, animal nests, beehives, and animal or plant products inside buildings. Larvae cause damage when they infest products and textiles such as wool and carpets, as well as museum specimens. Contact with larvae or shed larval exoskeletons can cause skin irritation.

Dianne Clarke

Hide Beetle ■ *Dermestes (Dermestinus) maculatus* TL 9mm (Dermestid Beetle)

DESCRIPTION Beetle typically dark brown to black, covered in fine hairs. Larvae dark brown, paler underneath, covered in hairs, with two upwards-curving, spine-like projections on final abdominal segment. **DISTRIBUTION** Introduced to Australia. Worldwide. **HABITAT AND HABITS** Typically associated with human habitation. Feeds on carrion and dry animal proteins, and usually breeds in such food sources, becoming a pest in products such as silk, meat and dog food. Often used in museums to clean skeletons of animal specimens. Its presence can aid forensic investigations in homicide or suicide cases.

Simon Grove

Dytiscidae (Predatory Diving Beetles)

Three-punctured Diving Beetle ■ *Cybister tripunctatus* TL 30mm

DESCRIPTION Smooth, streamlined, flattened oval body; dark green-brown to black with pale stripe around margin of body; long hindlegs, fringed with hairs. **DISTRIBUTION** Mainland Australia. Also southern Asia and Africa. **HABITAT AND HABITS** Typically inhabits slow-moving fresh water such as lakes and dams. Usually spotted as it surfaces to refresh the bubble of air it carries beneath its elytra. Adults and larvae are predatory, feeding on a range of small invertebrates. Adults can fly, and may be attracted to lights at night.

Dianne Clarke

Elateridae (Click Beetles)

Click beetles ■ *Monocrepidus spp. TL 10–20mm*

DESCRIPTION Elongated beetle, typically dark brown to reddish-brown. Small head partially hidden under pronotum, and longitudinal grooves on elytra. Larvae long, thin, sclerotized and yellow-brown in colour. **DISTRIBUTION** Throughout Australia. **HABITAT AND HABITS** Occurs in a variety of habitats. Biology of the species varies and for many is not well known; some predate on insect larvae, and others are reported to attack sugar cane. Larvae, called wireworms, live in rotting wood and soil, where they feed on plant matter; they can become agricultural pests.

Peter Rowland/kapeimages.com.au

Robert Webster (cc)

Black click beetles

■ *Melanotus* spp. TL 5–18mm

DESCRIPTION Elongated beetle, typically dark brown to black, with small head partially hidden under pronotum, and longitudinal grooves on elytra. Larvae long, thin, sclerotized and yellow-brown in colour. **DISTRIBUTION** Northern and eastern Australia. **HABITAT AND HABITS** Occurs in warm wet regions. Adults feed on pollen, and produce the characteristic clicking noise when trying to right themselves if flipped on their back. Larvae, called wireworms, live in moist, rich soils, where they feed on plant matter.

Scott Gilmore

Gyrinidae (Whirligig Beetles)

Whirligig Beetle

■ *Macrogyrus (Australogyrus) oblongus* TL 10mm

DESCRIPTION Flattened, streamlined, oval body, dark brown to greenish-black in colour; longitudinal ridges on elytra; forelegs longer than other legs. Each eye divided into halves. Larvae long, thin and pale bodied with dark head; feathery gills on abdomen. **DISTRIBUTION** Throughout Australia. **HABITAT AND HABITS** Inhabits fresh water, preferring calm shaded waters. Swims on surface in groups, sometimes in a characteristic rapid circular motion, and capable of diving and swimming underwater to avoid predators. Uses halved eyes to view above and below the water's surface simultaneously. Eats other organisms that fall on the water's surface, sensing the vibrations in the water. Larvae have gills and live at the bottom of waterbodies, predating on a range of soft-bodied invertebrates before crawling out of the water to pupate. Adults can fly, and may be attracted to lights at night.

LAMPYRIDAE (FIREFLIES)

Eastern fireflies ■ *Atyphella* spp. TL 5–10mm

DESCRIPTION Brown beetle with head hidden under extended pronotum, short antennae and light-emitting organ on underside of abdomen. Male has dark brown to black head, abdomen and elytra, and lighter brown pronotum. Female pale brown with reduced wings. **DISTRIBUTION** Northern and eastern mainland Australia. **HABITAT AND HABITS** Inhabits rainforests and damp forested areas, occasionally in gardens near bushland. Nocturnal, and usually seen in spring when males and females signal to each other. Light is produced by a biochemical reaction in the abdomen. Larvae live in moist leaf litter and are predatory on small invertebrates like snails and worms.

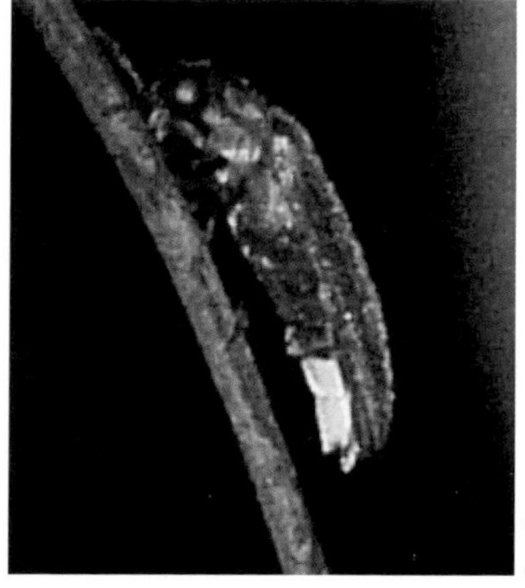

Peter Street

LUCANIDAE (STAG BEETLES)

Golden Green Stag Beetle ■ *Lamprima aurata* TL 35mm

DESCRIPTION Smooth, metallic iridescent beetle with pink-purple head. Females vary in colour from red-golden with green tinges to mostly blue-green. Males larger, with long, forwards-facing mandibles; inner surface with golden hairs; body typically golden-green with orange-red hues. **DISTRIBUTION** Eastern Australia. **HABITAT AND HABITS** Inhabits wet forested areas near coasts, and despite being widespread is uncommonly seen. Adults feed on flowers and fruits. Larvae live in rotting wood, feeding on the wood, fungi and other organic material. Males use their large mandibles to battle on fallen logs for mating rights with females; the losing beetle is usually thrown off the log.

John Nielsen

Lycidae (Lycid Beetles)

Net-winged beetles ■ *Porrostoma* spp. TL 10–20mm
(Lycid beetles)

DESCRIPTION Elongated black beetle with varying extents of brick-red, thick black legs, long, sometimes ornate antennae and longitudinal ridges on soft elytra. Larvae dark and heavily armoured. **DISTRIBUTION** Throughout Australia. Also NZ and Southeast Asia. **HABITAT AND HABITS** Adult beetles usually seen in forested areas flying during the day in warmer months. Adults short lived, and thought to feed on nectar and pollen as they are often found around flowers. Larvae typically inhabit wet leaf litter and moist logs. Lycid beetles contain chemicals that make them distasteful to predators, and rely on their bright aposematic colouration to ward off attackers. Many beetle species in other families mimic them, as do insects from other orders such as the Diptera, Hymenoptera and Lepidoptera.

Kristi Ellingsen

Red-shouldered Lycid ■ *Trichalus ampliatus* TL 20mm

DESCRIPTION Elongated black beetle with brick-red elytra, red 'shoulders' on sides of pronotum, thick black legs, and antennae and longitudinal ridges on soft elytra. **DISTRIBUTION** Eastern Qld and NSW. **HABITAT AND HABITS** Adult beetles usually seen in forested areas flying during the day in warmer months. Adults short lived, and thought to feed on nectar and pollen as they are often found around flowers. Larvae live in leaf litter and under bark, and thought to feed on organic matter. Lycid beetles contain chemicals that make them distasteful to predators, and rely on their bright aposematic colouration to ward off attackers. Many beetle species in other families mimic them, as do insects from other orders such as the Diptera, Hymenoptera and Lepidoptera.

Peter Rowland/kapeimages.com.au

Melyridae (Soft-winged Flower Beetles)

Red and Blue Pollen Beetle ■ *Dicranolaius bellulus* TL 8mm

DESCRIPTION Small, metallic blue and bright red beetle, with head and pronotum narrower than elytra. Tip of abdomen extends beyond elytra. Male has swollen third antennal segment. **DISTRIBUTION** Throughout Australia. **HABITAT AND HABITS** Occurs in a variety of habitats. Eats pollen, as the name suggests, but is also predatory, feeding on eggs, larvae and small slow-moving arthropods. Used as biological control generally for lepidopteran pests of crops and gardens. Larvae live in soil.

Donald Hobern (cc)

Passalidae (Bess Beetles)

Black Bess beetles ■ *Mastachilus* spp. TL 30–40mm

DESCRIPTION Typically shiny dark brown to black cylindrical beetle, with wide head, thick jaws, broad pronotum, narrow 'waist' between pronotum and elytra, and longitudinal grooves on elytra. Sometimes has light yellow hairs on thorax and abdomen. Larvae long, pale and soft bodied, with pale brown head and only 4 functional legs. **DISTRIBUTION** Eastern Australia. Also PNG. **HABITAT AND HABITS** Occurs in relatively undisturbed forested areas. Unlike other beetles it is subsocial, living in small groups usually in rotting logs or tree stumps. Cares for young grubs by preparing food and helping construct pupal cases. Larvae and adults communicate using different sounds, produced by rubbing legs together in larvae, and in adults rubbing abdomen against hindwings.

Dianne Clarke

SCARABEIDAE (SCARAB BEETLES)

Christmas Beetle ■ *Anoplognathus porosus* TL 25mm

DESCRIPTION Several species in the *Anoplognathus* genus are known as Christmas Beetles. All are similar in appearance, being metallic orange-brown to pale gold in colour, with variable dark spotting, some with greenish tinges. Larvae pale and soft bodied with darker sclerotized head. **DISTRIBUTION** Eastern Australia. **HABITAT AND HABITS** Found in wooded habitats, including dry sclerophyll forests, woodland and household gardens. Adults feed on leaves of various trees, such as eucalypts and melaleucas. Eggs laid in soil and larvae (known as curl grubs) feed on plant roots or decomposing vegetation for about 1–2 years. When fully grown larvae pupate in soil and emerge in warmer months (mainly November–January) as adults. Large numbers of adults can be attracted to artificial lights around houses.

Rachel Whitlock

Green scarabs ■ *Diphucephala spp. TL 10mm*

DESCRIPTION Small, metallic green beetle with reddish or golden tinges. Short hairs on thorax, abdomen and legs, and elytra often with pitted appearance. Larvae pale and soft bodied, with darker sclerotized head. **DISTRIBUTION** Throughout Australia. **HABITAT AND HABITS** Occurs in a variety of habitats, from eucalypt woodland to urban gardens. Feeds on leaves and sap from a wide variety of plants such as *Melaleuca*, *Leptospermum*, *Acacia* and *Alphitonia*. Also feeds on nectar and pollen from flowers. Larvae live in soil, feeding on plant roots and organic matter. Large populations can cause significant defoliation to orchards and plantation trees.

Kristi Ellingsen

Fiddler Beetle ■ *Eupoecila australasiae* TL 20mm

DESCRIPTION Beetle dark brown to black with bright yellow to lime-green markings, and reddish-brown legs. Larvae pale and soft bodied with small brown head. **DISTRIBUTION** Eastern Australia. **HABITAT AND HABITS** Typically occurs in woodland, heathland, parks and gardens. Strong flier, moving from plant to plant, feeding on plant sap and nectar and pollen of flowers from trees such as *Angophora*, *Backhousia* and *Melaleuca*. Larva lives in rotting wood or soil rich with organic matter, which it eats until it pupates, emerging as beetle in early summer. Common name refers to fiddle or violin-like markings on its back.

John Tann (cc)

Black Lawn Beetle ■ *Heteronychus arator* TL 15mm

(African Black Beetle, Black Beetle)

DESCRIPTION Shiny black, rounded beetle. Larvae pale and soft bodied with brown sclerotized head. **DISTRIBUTION** Introduced to Australia. Tropical and subtropical regions worldwide; native to southern Africa. **HABITAT AND HABITS** Occurs in coastal regions near main cities and agricultural activity. Larvae and adults burrow and feed on plant roots and tubers underground. Adult beetle also chews on stems at or under soil surface, and can fly to new areas. In gardens can cause damage to lawns, vegetables and ornamental plants. Due to its generalist diet it is an agricultural pest of pasture, grains, sugar cane, grapevines, potatoes and newly planted eucalypts, among others.

Peter Rowland/kapeimages.com.au

Native dung beetles ■ *Onthophagus* spp. TL 2–11mm

(Scarab beetles)

DESCRIPTION Small, usually shiny, dark-coloured, rounded beetle, with broad, shovel-like head and pronotum and small, fan-like antennae; may have hairs on underside of body. Some species have horns on head and pronotum in males. Larvae pale and soft bodied with dark sclerotized head. **DISTRIBUTION** Throughout Australia. Worldwide. **HABITAT AND HABITS** Found in a variety of habitats, usually wherever large animals occur to provide food for larvae. Typical of dung beetles, larvae are provided with animal dung to feed on, though some species feed on carrion or fungi instead. Large genus with more than 200 species in Australia, some introduced, and over 1,500 species worldwide.

CSIRO ScienceImage (cc)

Mottled Flower Scarab ■ *Protaetia fusca* TL 17mm

(Mango Flower Beetle)

DESCRIPTION Adult dark brown to black, occasionally reddish or greenish, with patchy, mottled white markings. Larva pale and soft-bodied with dark sclerotized head. **DISTRIBUTION** Recorded in most Australian states, but possibly absent from cooler regions in the south. Native to Southeast Asia, introduced in warmer regions around the world. **HABITAT AND HABITS** Adult beetles can be found year-round, typically in forested areas or orchards in wetter regions. It is active during the day; feeds on flower nectar, pollen, fruit and sap of a variety of trees. Roses, avocados, mangoes and peaches are among the plants it feeds on in Australia, and in large numbers it may become an agricultural pest. Also recorded raiding European honeybee hives for honey. The larva lives in rotting wood and soil rich in decomposing organic matter.

Peter Rowland/kapeimages.com.au

Rhinoceros Beetle ■ *Xylotrupes australicus* TL 70mm

(Elephant Beetle, Coconut Palm Beetle)

DESCRIPTION Large, shiny, dark brown to black beetle; male has horn-like projections on head and pronotum. Larva a large pale grub, with thick, soft body and reddish head. **DISTRIBUTION** Eastern Qld and north-eastern NSW. Also Southeast Asia. **HABITAT AND HABITS** Inhabits rainforest areas near coasts, where adult beetles feed on nectar and rotting fruit. Typically seen during wet season, and often attracted to lights at night. Males use their large horns to battle each other, and the losing beetle may be flipped on to its back and unable to right itself. Capable of making a hissing or squeaking noise if disturbed. Larvae live in soil rich in decomposing organic matter, and sometimes found in garden compost heaps.

Donald Hobern (cc)

Staphylinidae (Rove Beetles)

Rove beetles ■ *Paederus* spp. TL 10mm

DESCRIPTION Small, elongated beetle, usually with contrasting black and bright orange-red patches, and short elytra only covering part of the soft body. **DISTRIBUTION** Eastern, southern and south-western Australia. **HABITAT AND HABITS** Inhabits moist areas usually near water sources, and can migrate to drier areas during floods or heavy rain. Active during the day, though often attracted to lights at night. Larvae and adult beetles are predatory. Some *Paederus* species contain toxins, so bright colouration is likely to be aposematic. Toxins cause irritation and blistering if an insect is crushed against the skin; affected area should be thoroughly washed with soapy water immediately after contact.

Reiner Richter

Tenebrionidae (Darkling Beetles)

Honeybrown Darkling Beetle ■ *Ecnolagria grandis* TL 20mm

(Copper Kettle Beetle)

DESCRIPTION Beetle shiny golden-brown to darker reddish-brown body, covered in short hairs giving fuzzy appearance. Head and pronotum narrower than oblong-shaped elytra. Larvae dark brown, also covered in fine hairs. **DISTRIBUTION** Southern and eastern Australia. **HABITAT AND HABITS** Occurs in wooded areas such as sclerophyll forests, parks and gardens, usually in moist areas or areas with decent rainfall. Often found resting on plant leaves or grasses, which it eats. Larvae found on forest floor among leaf litter.

Andrew Allen

Pie-dish Beetle ■ *Pterohelaeus* spp. TL 10–35mm

DESCRIPTION Adult beetle typically black to bluish, with longitudinal ridges down elytra, and margins of body flattened to form a 'pie-dish' shape. Larvae yellow-brown, elongated grubs. **DISTRIBUTION** Throughout Australia. **HABITAT AND HABITS** Occurs in a variety of habitats, typically in forested areas. Shelters during the day, usually on the ground or under tree bark, and emerges at night to forage for food. Larva burrows in soil and feeds on plant roots. Its distinctive body shape is thought to help the beetle fit in hiding places and to provide defence against predators such as scorpions.

Peter Rowland/kapeimages.com.au

Mealworm Beetle ■ *Tenebrio molitor* TL 20mm
(Yellow Mealworm)

DESCRIPTION Adult beetle black, with wide, shield-like pronotum, longitudinal ridges on elytra, yellow-brown abdomen and clear hindwings. Larva a yellow-brown, elongated grub, darker near head and tip of abdomen. **DISTRIBUTION** Introduced to Australia. Worldwide; thought to originate in Europe. **HABITAT AND HABITS** Like in many other countries worldwide, commercially bred in Australia, with larvae sold as bait or as feeder insects for pets like reptiles and birds. Also used in biological research, as a food source and even kept as pets. Scavenger, feeding on dead arthropods, meat scraps and stored grain; may become a pest in grain-storage facilities.

Rachel Whitlock

Asilidae (Robber Flies)

Giant Yellow Robber Fly
■ *Blepharotes coriarius* TL 45mm

DESCRIPTION Large, grey and black, with bright yellow hairs on upper surface of abdomen and greyish-brown wings extending beyond tip of abdomen. **DISTRIBUTION** Recorded over much of Australia, including Qld, NSW, Vic, SA and southern WA. **HABITAT AND HABITS** Occurs in forests, woodland and gardens, where its buzzing flight often betrays its presence when not perched on side of tree, branch or vegetation. Preys on other insects, including beetles and other flies, typically caught in mid-air using its strong claws. Helpless victim is then pierced with proboscis and injected with toxic saliva and digestive enzymes before body contents are sucked out.

Peter Rowland/kapeimages.com.au

Peter Rowland/kapeimages.com.au

Robber Fly

■ *Colepia ingloria* TL 30mm

DESCRIPTION Large, grey and brown, with reddish legs and greyish-brown wings not extending beyond tip of abdomen. **DISTRIBUTION** Eastern Australia, from north-eastern Qld to eastern Vic. **HABITAT AND HABITS** Occurs in forests, woodland and gardens, where its buzzing flight often betrays its presence. At rest perches on tree trunks, branches, vegetation or on the ground. Preys on other insects, including beetles and other flies, typically caught in mid-air using its strong claws. Helpless victim is then pierced with proboscis and injected with toxic saliva and digestive enzymes before body contents are sucked out.

CALLIPHORIDAE (BLOWFLIES)

Lesser Brown Blowfly ■ *Calliphora (Paracalliphora) augur* TL 6–11mm

DESCRIPTION Mostly golden-brown with clouded metallic blue thorax and central shield on abdomen, and distinctive bristles on lower sides of body. Forewings membranous, and hindwings reduced to club-like halteres. Eyes red and almost meet in male, but with clear separation in female. **DISTRIBUTION** Widespread throughout Australia. **HABITAT AND HABITS** Occurs in open woodland, mallee and open pastures. Strongly associated with sheep, and commonly enters houses during winter, when numbers are highest. Breeding can occur year round, and female lays eggs or live young (sometimes both) in flesh of carcasses, in open wounds of injured animals or in wool of sheep.

Peter Rowland/kapeimages.com.au

Oriental Latrine Fly ■ *Chrysomya megacephala* TL 11mm

DESCRIPTION Metallic blue-green body and large red eyes that meet in male, but are separated in female, and sponge-like mouthparts. Single pair of large, transparent wings; second pair has been modified into stabilizing appendages called halteres. **DISTRIBUTION** Throughout Australasia, and most of the world. In Australia, has been recorded in every state and territory except Tas. **HABITAT AND HABITS** Lives in close association with human habitation, and able to thrive in warm climates with increased breeding success, but also lives successfully in cool areas. Adults attracted to human refuse, rotting carcasses and faeces, and recorded as carrying several types of disease-causing bacteria, including *Staphylococci typhi*, *Streptococcus aureus*, *Salmonella typhi*, *Pseudomonas aeruginosa* and *Bacillus* spp.

Jenny Thynne

CERATOPOGONIDAE (BITING MIDGES)

Biting midges ■ *Culicoides* spp. TL 3mm

(Sandflies)

DESCRIPTION Tiny, and the smallest of all flies that feed on blood. Adults resemble mosquitoes, with wings held in 'V' shape while feeding or at rest. **DISTRIBUTION** Throughout Australia in suitable habitats. **HABITAT AND HABITS** Found mainly in coastal areas, including swamps, tidal flats, creeks, lagoons and mangroves, and most active at dawn and dusk. Mainly feeds on nectar and other plant exudates, but females readily attack exposed skin of humans to obtain blood. Victim is often not aware of being bitten until swelling, itchiness and redness appear. Sensitive people can exhibit severe local allergic reactions, including sores that persist and weep for weeks.

CSIRO ScienceImage (cc)

CHIRONOMIDAE (NON-BITING MIDGES)

Peter Rowland/kapeimages.com.au

Non-biting midges

■ Multiple species TL to 20mm

DESCRIPTION Members of large family superficially similar, typified by thin body and long, thin legs. Wings membranous. **DISTRIBUTION** Widely distributed throughout Australia and most of the world. **HABITAT AND HABITS** Found in and around water, including rivers, still lakes, dams, ponds and temporary puddles. Adults attracted to lights and often congregate around house lights and on windows. Larvae aquatic, living around submerged vegetation, in benthic debris or in open water. Some species can tolerate highly polluted water. Adults have short lifespans, existing solely to breed and feeding off energy stored in their body for survival, although they may also feed occasionally on nectar. Larvae of different species feed on algae, organic detritus, diatoms, macrophytes, oligochaetes, roundworms and small invertebrates.

CULICIDAE (MOSQUITOES)

Striped Mosquito ■ *Aedes (Rampamyia) notoscriptus* TL 4 mm

DESCRIPTION Dark greyish with conspicuous white to yellowish, lyre-shaped lines on dorsal shield (scutum). Legs have pale bands, and proboscis has distinct white central band. **DISTRIBUTION** Throughout coastal and inland mainland Australia, including Tas. **HABITAT AND HABITS** Requires small pools of fresh water, including tree hollows, garden ponds, pot plants, bird baths and gutters, for larval development, and adults travel up to 0.4km from the site they develop in. Readily attacks at dawn and dusk, but also in shaded areas during the day, and is a major pest in urban gardens. Key vector of dog heartworm, and known vector of Ross River virus and Barmah Forest virus; also found to be able to carry Murray River encephalitis virus in laboratory tests.

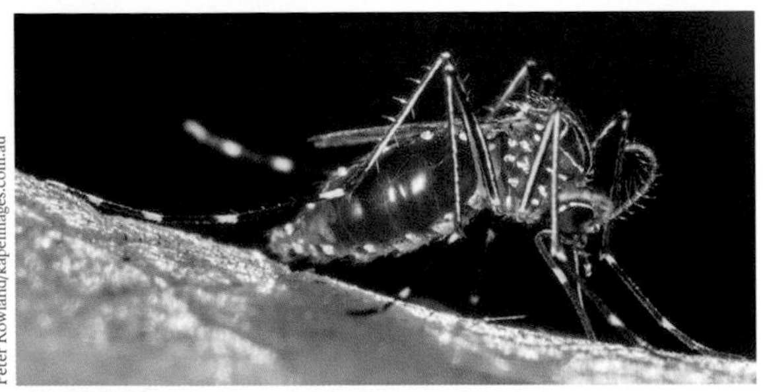
Peter Rowland/kapeimages.com.au

Yellow Fever Mosquito ■ *Aedes (Stegomyia) aegypti* TL 3.8 mm

DESCRIPTION Generally dark with white, lyre-shaped pattern above, and white patches on abdomen, sometimes forming continuous bands. Legs have pale and white bands. **DISTRIBUTION** Introduced into Australia and formerly widespread in WA, NT, Qld and NSW, but now considered restricted to northern and eastern Qld. Native to Africa. **HABITAT AND HABITS** Found around artificial water sources, including water tanks, buckets, bird baths and underground pits, which are needed for development of aquatic young. Most active in shaded areas during the day and early evening. The only known vector of Dengue fever virus in Australia, but also capable of transmitting Ross River virus and Murray encephalitis.

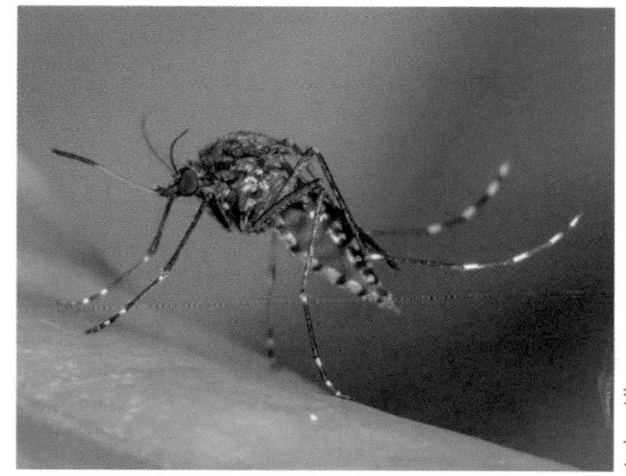

Andrew Allen

Predatory Mosquito ■ *Toxorhynchites (Toxorhynchites) speciosus TL 12mm* (Elephant Mosquito)

DESCRIPTION Adults large, dark and shiny, with strongly backwards-curved proboscis. Abdomen has large paler patches and hairy tufts at rear, and legs have pale subterminal bands. Male has bushy antennae. **DISTRIBUTION** Coastal eastern Australia from southern Cape York Peninsula, Qld, to around Sydney, NSW. **HABITAT AND HABITS** Inhabits a range of vegetated areas containing waterbodies, which are required for larval development, and regularly enters gardens. Adults feed on nectar, sap, and other plant juices. Female does not require a blood meal for egg production, so is not attracted to humans and does not bite them. Larvae prey on other mosquito larvae, including their own species.

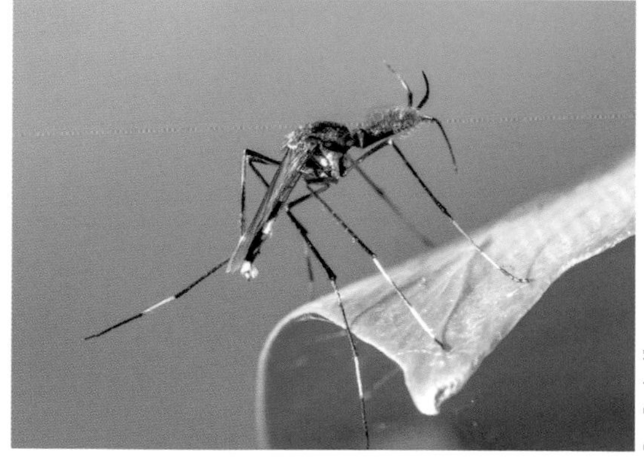

Peter Rowland/kapeimages.com.au

DOLICHOPODIDAE (LONG-LEGGED FLIES)

Long-legged flies ■ Multiple species TL 1–9mm

DESCRIPTION Characterized by long, slender legs. Most species metallic green or bronze, occasionally yellow, and banded with black, but a few can be dull brown or black. Abdomen often slender and wings long, narrow and membranous, with reduced venation and varying black markings. Eyes usually green or red. **DISTRIBUTION** More than 7,000 species known worldwide, including over 1,400 in Australasia/Oceania, with many more to be formally described. Representatives widespread in Australia. **HABITAT AND HABITS** Found in a variety of habitats. Commonly seen walking on foliage, tree trunks, rocks, moist ground and occasionally surfaces of still ponds. Adults prey on soft-bodied invertebrates, including aphids, mites and thrips. Larvae hunt and scavenge under bark, or in ground litter or soil, for other insect larvae.

Peter Rowland/kapeimages.com.au

DROSOPHILIDAE (FRUIT FLIES)

Fruit flies ■ *Drosophila* spp. TL 2–3mm
(Vinegar Flies)

Kristi Ellingsen

DESCRIPTION Typically yellowish- or reddish-brown to black, with sparse blackish bristles and membranous forewings. Eyes red and antennae very short with enlarged tips. **DISTRIBUTION** About 1,500 species distributed in tropical and temperate regions worldwide. **HABITAT AND HABITS** Found in a wide range of habitats, including arid deserts, alpine areas, rainforests, woodland and gardens. Some species, such as *D. melanogaster*, commonly encountered in houses and commercial premises. Typically seen around fruits, particularly when they are over-ripe or decomposing, fungi or fleshy flowers, in which larvae feed after hatching from eggs. Males defend territories around breeding sites. Males of *D. bifurca* have the longest sperm of any studied organism, up to 58mm, or 20 times the body length.

Lauxaniidae (Lauxaniid Flies)

Dotted-wing lauxaniid flies ■ *Homoneura* spp. TL 6mm

DESCRIPTION Overall colouration brownish to orange with reddish compound eyes. Body has sparse blackish hairs, and legs moderately long. Wings longer than body, translucent smoky-grey, and with pattern of darker spots. Some members of genus can be partially or entirely black and have no dotted wing pattern. **DISTRIBUTION** Eastern, south-eastern and south-western Australia; probably more widespread than records indicate. More widely distributed throughout the world. **HABITAT AND HABITS** Occurs in forests, woodland, heaths and mangroves, typically preferring moist, shaded undergrowth or damp soil. Seen less commonly in grassland and similarly open habitats. Larvae saprophagous, feeding on decaying organic matter and fungi in leaf litter and similar vegetation.

Peter Rowland/kape.mages.com.au

Muscidae (House Flies)

House Fly ■ *Musca domestica* TL 8mm

DESCRIPTION Thorax and abdomen greyish (although abdomen is paler), with 4 blackish longitudinal lines on thorax and single pair of transparent, triangular wings. Eyes reddish and mouthparts sponge-like. **DISTRIBUTION** Throughout most of Australia and the rest of the world. **HABITAT AND HABITS** Abundant in urban and rural areas, wherever suitable breeding sites (human and animal waste) are found. Will investigate human foodstuff, household rubbish, septic waste and rotting vegetable and animal matter, using sponge-like mouthparts to suck up liquids and regurgitate saliva on to more solid foods to break them down. In doing so, transfers microscopic organisms between these sites as it travels, which can lead to diseases such as salmonella, dysentery, hepatitis, cholera, poliomyelitis and typhoid fever.

Peter Rowland/kapeimages.com.au

Australian Bush Fly ■ *Musca vetustissima* TL 7mm

DESCRIPTION Ashy-grey with 2 diverging black stripes on thorax, becoming 4 near head, and red eyes. The 2 rounded, triangular, transparent wings touch or overlap slightly when at rest. Abdomen yellowish and eyes touching in male; abdomen grey and black, and eyes separated in female. **DISTRIBUTION** Endemic to Australia, where it is widespread in drier areas of all states and territories, except Tas. **HABITAT AND HABITS** Bothersome fly during summer months, with swarms of flies landing on backs of animals (including humans) and crawling around eyes, nose, ears and mouth, searching for moisture and proteins (female). Responsible for spreading bacteria such as *Escherichia coli* and *Salmonella*.

Thomas Rowland/kapeimages.com.au

Stable Fly ■ *Stomoxys calcitrans* TL 7 mm

DESCRIPTION Thorax and abdomen greyish; darker longitudinal lines on thorax and pale area between these. The 2 transparent wings are held widely apart when not in flight. Piercing and sucking mouthparts. **DISTRIBUTION** Warmer parts of Australia, particularly where livestock industries are located. **HABITAT AND HABITS** Found mainly in rural areas and on beaches, and seldom enters urban locations, except where livestock and horse stabling occur. Both sexes drink blood, and can consume up to 3 times their body weight. Bites are painful and can produce itchiness, urticaria and cellulitis. Allergic reactions can also occur in some individuals, accompanied by wheezing and hives.

Dianne Clarke

Neriidae (Banana-stalk and Cactus Flies)

Banana-stalk Fly ■ *Telostylinus lineolatus* TL 20mm

DESCRIPTION Dark brown and yellowish-cream, with large red eyes on elongated head, and moderately long, thickened antennae with single long, thin arista arising from each (used to detect heat and moisture). Legs long and stilt-like, and wings darkened, long and membranous. **DISTRIBUTION** Along coastal eastern Australia, from northern Qld south to around Wollongong, NSW, and throughout Indo-Pacific. **HABITAT AND HABITS** Found in various vegetated areas, including gardens, where it aggregates on rotting timber, flowers and fruits. Males maintain territories around egg-laying sites and jostle with each other by raising up and interlocking antennae. Their forelegs have spines and are used to strike their opponents and to attempt to get them in a headlock. Adults and larvae feed on rotting vegetable matter.

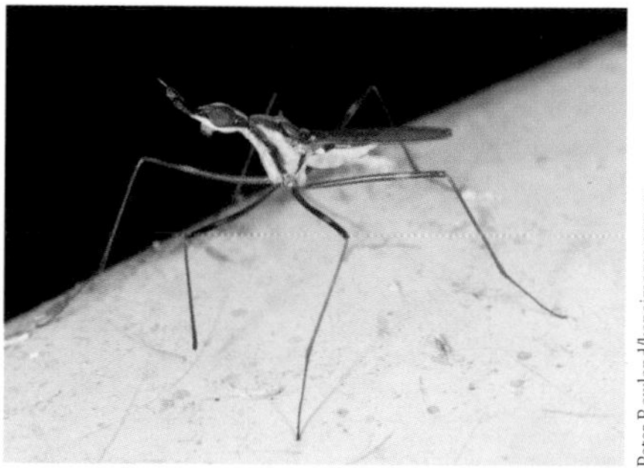

Peter Rowland/kapeimages.com.au

Platystomatidae (Signal Flies)

Boatman Fly ■ *Pogonortalis doclea* TL 5mm

DESCRIPTION Adults metallic greenish and black, with black bristles and white hairs, giving a dusty appearance. Wings transparent with some darker markings, typically held outstretched to sides and waved like oars. Eyes red and with clear separation between them and bordered by thin white lines. **DISTRIBUTION** Through eastern and southern Australia, including Tas. Introduced overseas, including to Indonesia and USA. **HABITAT AND HABITS** Occurs in open grassland and lightly wooded areas, where it is attracted to fresh mammal dung for feeding. Short-lived species, but active year round and females produce 2 or more broods of offspring per year. Males territorial around egg-laying sites, using broadened head and thickened lower facial bristles in head-to-head combat against rivals.

Andrew Allen

Sarcophagidae (Flesh Flies)

Flesh Flies ■ *Sarcophaga* spp. TL up to 20mm

DESCRIPTION Numerous species with similar appearance. Typically grey to yellowish thorax and head, bluish-grey abdomen and often bright red eyes. Top of thorax often marked with 3 longitudinal dark stripes and abdomen chequered. Forewings membranous with darker veins, and hindwings reduced to stabilizing halteres. Numerous long black bristles at rear of abdomen. **DISTRIBUTION** Widely distributed throughout Australia. **HABITAT AND HABITS** Occurs in woodland, grassland and urban areas, where adults occasionally feed from flowers. Adults arrive at carcasses shortly after blowflies, and females lay live larvae (maggots) in rotting corpses, the eggs they carry hatching in the uterus before laying to allow larvae to feed immediately.

Peter Rowland/kapeimages.com.au

Stratiomyidae (Soldier Flies)

Black Soldier Fly ■ *Hermetia illucens* TL 15–20mm

(American Soldier Fly)

DESCRIPTION Mostly black with some greyish-white banding on abdomen and varying amount of blue, green and red metallic sheen on upper and lower body. Translucent patches on second last tergite, wings smoky-brown and membranous, and legs predominantly white. Antennae about twice length of head. **DISTRIBUTION** Probably introduced and now found throughout Australia, but possibly absent from SA and Tas. Has spread to every continent, originating in Americas. **HABITAT AND HABITS** Occurs in a wide variety of tropical and subtropical habitats. Strongly associated with decomposing organic matter, including human remains and manure, adults typically arriving in more advanced stages to lay up to 800 eggs. After hatching, larvae feed on decaying matter for up to 14 days, then pupate into adult form, which lives up to 8 days.

Dianne Clarke

SYRPHIDAE (HOVERFLIES)

Common Hoverfly ■ *Melangyna (Austrosyrphus) viridiceps* TL 10–15mm

Kristi Ellingsen

DESCRIPTION Abdomen black with incomplete yellow bands, and dorsoventrally flattened. Distinguished from the Yellow-shouldered Stout Hoverfly (see below) by entirely black thorax. Eyes very large and compound, larger and close to touching in males. Wings transparent. **DISTRIBUTION** Widespread in eastern, southern and south-western Australia, including Tas. **HABITAT AND HABITS** Found in various habitats where adults are commonly seen hovering near flowers, alighting at times to feed on pollen and nectar. Larvae voraciously feed on aphids, consuming about 300 during this life-cycle stage, and other small insects such as thrips and scale insects.

Yellow-shouldered Stout Hoverfly

■ *Simosyrphus grandicornis* TL 10–15mm
(Common Hoverfly)

DESCRIPTION Body yellow, with black, crescent-shaped barring on upper surface of abdomen, forming large yellow spots, and black on upper thorax. Eyes very large and compound, larger and close to touching in male, and antennae yellow. Wings transparent, with false margin on rear edge. **DISTRIBUTION** Widespread in Australia, where it is native. Also introduced to many countries throughout Pacific. **HABITAT AND HABITS** Common and widespread, and particularly abundant in areas with flowering plants, where adults most often seen hovering near flowers, alighting at times to feed on pollen and nectar. Larvae feed on aphids and other soft-bodied arthropods.

Peter Rowland/kapeimages.com.au

Tabanidae (Horse and March Flies)

Peter Rowland/kapeimages.com.au

March Fly ■ *Tabanus australicus* TL 13mm

DESCRIPTION Solidly built with large, reflective, iridescent eyes that meet in middle in male. Eyes larger than those of the Stable Fly (see p. 110), and green in female and reddish-brown in male. **DISTRIBUTION** Warmer parts of Australia, including Qld, NSW, Vic, WA and NT. **HABITAT AND HABITS** Favours moist forests and woodland, particularly in vicinity of water. Females drink blood to provide protein for their eggs to develop, slicing through a victim's skin with sharp mouthparts. Males feed on nectar and other plant secretions. Bites painful and can produce itchiness, lesions, urticaria, cellulitis and fever. Allergic reactions can also occur in some individuals, with symptoms including wheezing, hives, muscle weakness and potential anaphylaxis.

Tachinidae (Parasitic Flies)

Tachinids ■ *Rutilia* spp. TL up to 20mm
(Parasitic flies)

DESCRIPTION Adult flies brown to metallic green, sparsely covered with short black bristles and shorter fine white hairs. Eyes dull red and compound, and wings membranous with darker veins. **DISTRIBUTION** Large genus native to Australia and Asian region. In Australia most commonly seen in eastern and southern states. **HABITAT AND HABITS** Found in vegetated habitats, including forests, woodland, grassland, wetlands and urban gardens, occasionally entering buildings. Adults parasitize larvae of scarab beetles. Females normally deposits eggs on soil, and larvae burrow in search of host after hatching. A larva then pupates in soil or ground litter after the third instar.

Kristi Ellingsen

TEPHRITIDAE (FRUIT FLIES)

Queensland Fruit Fly ■ *Bactrocera (Bactrocera) tryoni* TL 5–8mm (Q-fly)

Peter Rowland/kapeimages.com.au

DESCRIPTION Reddish-brown to dark brown and yellow, with red eyes and membranous wings. **DISTRIBUTION** Mainly through Qld and NSW, but also in major fruit-growing areas of Vic and SA. **HABITAT AND HABITS** Found in most areas with fruiting trees, including commercial orchards, where adults feed on nectar, fruit juices and honeydew. Adults breed late spring to autumn, and able to survive during winter months. Females lay eggs in fruits, either by oviposition through skin, or into open wounds and puncture marks from other fruit pests. Eggs hatch after 2–3 days and larvae feed on flesh of fruit for about 5 days before falling to the ground to pupate in soil. Major horticultural pest of many fruit species, with strict controls in place to prevent its spread.

TIPULIDAE (CRANE FLIES)

Crane flies ■ *Leptotarsus* spp. TL 15–30mm

DESCRIPTION Often mistaken for large mosquito. Body long and narrow, mostly orange with black longitudinal markings on abdomen, and legs very long and thin. Abdomen of female swollen when carrying eggs, and with long, thin ovipositor. Wings longer than body, translucent brownish with iridescent sheen. **DISTRIBUTION** Throughout Australia, largely absent from central Australia. **HABITAT AND HABITS** Found in wet areas of woodland and grassland, as well as swamps, where adults can be seen feeding on nectar or drinking from droplets of water, or resting on foliage, often in small groups. Larvae live in soil or decaying organic matter.

Andrew Allen

Tiger Crane Fly

■ *Nephrotoma australasiae* TL 15mm

DESCRIPTION Body yellow to orange with black bands on abdomen and black markings on thorax. Legs very long and slender, and mostly black; wings smokey-grey with iridescent sheen and darker venation. Female typically has larger abdomen when carrying eggs, and long ovipositor. **DISTRIBUTION** Eastern Australia, mainly along Qld coast, and south-western WA. **HABITAT AND HABITS** Found in forests and woodland, including gardens, where adults can be seen resting on leaves, or occasionally feeding on nectar or drinking droplets of water in shaded areas near water. Hairs on legs and wings arranged in a manner to repel water. Short lived as adults, living solely to breed and lay eggs. Larvae feed on plant roots.

Rachel Whitlock

BITTACIDAE (HANGINGFLIES)

Hangingfly ■ *Harpobittacus australis* TL 20mm
(Scorpionfly)

John Nielsen

DESCRIPTION Adult body orange, black bands on long legs, black eyes, narrow wings and elongated jaws. **DISTRIBUTION** South-eastern Australia. **HABITAT AND HABITS** Adult typically seen hanging from low vegetation in moist habitats, waiting to ambush prey. Larvae resemble caterpillars with developed mandibles and no prolegs, and may be seen in leaf litter and low vegetation, where they feed on decaying organic matter.

Choristidae (Scorpionflies)

Autumn Scorpionfly

■ *Chorista australis* TL 20mm

John Tann (cc)

DESCRIPTION Adult body black and orange, head orange with black eyes and antennae, and wings transparent orange with black veins. **DISTRIBUTION** South-eastern Australia. **HABITAT AND HABITS** Found in moist areas on herbaceous plants, low shrubs and occasionally on grasses. Usually seen perched on or hanging from vegetation by front legs, waiting to seize prey using raptorial hindlegs. Prey includes soft-bodied insects and spiders. Male secretes pheromone to attract female and passes prey item to her at onset of copulation. Larvae resemble caterpillars with developed mandibles and no prolegs, and are usually found in moss or soft soil, or under debris.

Pulicidae (Fleas)

Dog Flea ■ *Ctenocephalides canis* TL 4mm

DESCRIPTION Reddish-brown to black. Wingless, laterally compressed body with spiny head and large abdomen. **DISTRIBUTION** Worldwide. Thought to originate in Africa. **HABITAT AND HABITS** Primary hosts are domestic and wild dogs, foxes and other canids, though it often infests cats and occasionally humans. Bites can cause allergic reactions, and it is a vector for tapeworm and disease-causing bacteria. Historical records of it occurring in Australia, but no confirmed reports in recent years. The similar Cat Flea (see p. 118) is commonly found on domestic dogs and other pets.

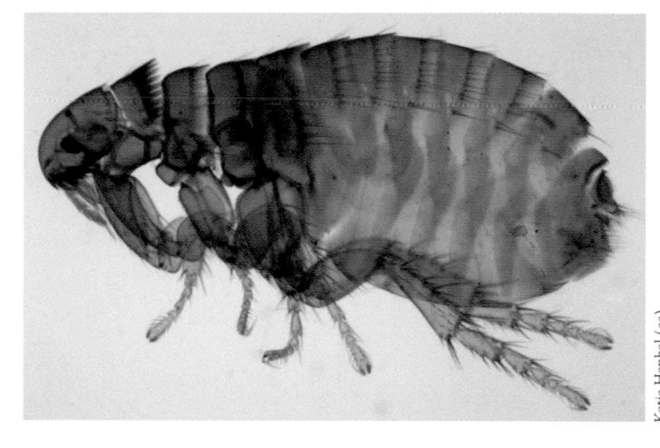

Karja Henbel (cc)

Cat Flea ■ *Ctenocephalides felis* TL 4mm

DESCRIPTION Reddish-brown to black, wingless, laterally compressed body, with spiny head and large abdomen. **DISTRIBUTION** Worldwide. Thought to originate in Africa. **HABITAT AND HABITS** The most commonly encountered flea in Australia, widespread in homes and urban areas across the country. Primarily feeds on blood of domestic cats and dogs, and also other mammals such as rats, possums and sometimes humans. Bites can cause allergic reactions, and it is a vector for parasites such as tapeworm and disease-causing bacteria. The various life stages may be found in bedding, clothing and other surfaces that infested hosts come into contact with.

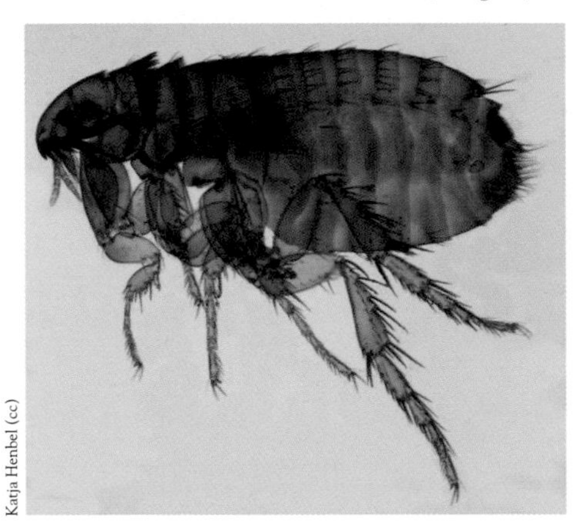

Katja Henbel (cc)

Human Flea ■ *Pulex (Pulex) irritans* TL 3.5mm

(House Flea)

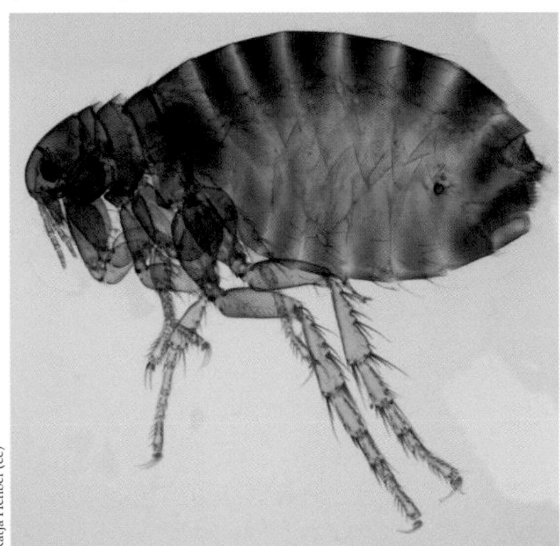

Katja Henbel (cc)

DESCRIPTION Reddish-brown, wingless, laterally compressed body. **DISTRIBUTION** Worldwide, thriving in temperate regions. Probably originated in South America. **HABITAT AND HABITS** Occurs primarily in urban areas. Feeds on a wide variety of mammal hosts, from humans to rodents and bats, and even bites birds such as chickens. Bites can trigger allergies, and it can carry parasites, bacteria and viral pathogens. Like the Dog Flea (see p. 117), this species is relatively rare in Australia.

Oriental Rat Flea

■ *Xenopsylla cheopis* TL 4mm

DESCRIPTION Dark reddish-brownish, laterally compressed, wingless body and hard exoskeleton. **DISTRIBUTION** Worldwide in tropical, subtropical and some temperate regions. **HABITAT AND HABITS** Widespread in urban areas, and in homes found in clothing and bedding near sleeping areas of human hosts. Piercing and sucking mouthparts used to feed on blood of host animals, normally rats, but can transmit murine typhus (or rat-flea typhus) to humans when its droppings enter the bloodstream. Vector for the bacterium *Yersinia pestis*, which was transferred from rats to humans, causing Europe's well-known bubonic plague of the Middle Ages.

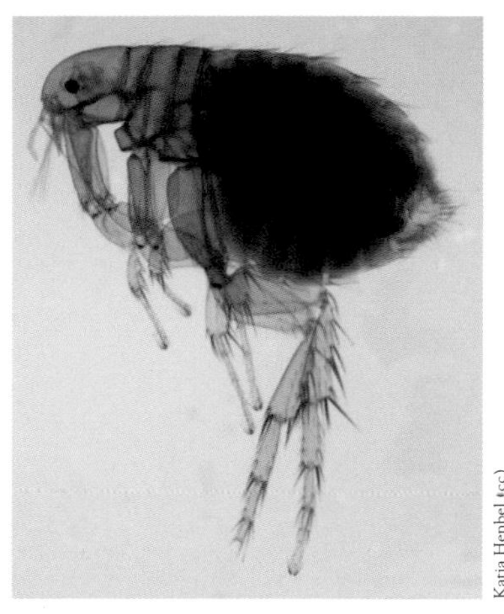

Katja Henbel (cc)

ANTHELIDAE (AUSTRALIAN WOOLLY BEARS)

Common Anthelid ■ *Anthela acuta* WS 30–50mm

DESCRIPTION Moths variable, but typically buff with pair of darker spots on each forewing, line through centre of each wing, and line of smaller spots along edges of rear wings. Female larger than male. Caterpillars brown and hairy, with pairs of white-and-pink spots, and longer tufts of pale brown to whitish hairs. **DISTRIBUTION** Eastern Australia. Most individuals collected in south-eastern Qld, eastern NSW, Vic and Tas. **HABITAT AND HABITS** Found in a wide range of habitats. Caterpillars feed mainly at night on a variety of vegetation, depending on the race and distribution. When fully grown, at *c.* 50mm, caterpillar constructs a hairy cocoon in an outer wall of leaves, placed on the ground in debris. Long hairs can cause mechanical skin irritations.

Dianne Clarke

Urticating Anthelid ■ *Anthela nicothoe* WS 70–100mm

DESCRIPTION Moths grey to orange-brown, with darker wavy lines across wings and pair of paler spots on each forewing. Female larger than male. Caterpillars brown and hairy, with body segments having small, hairless gaps separating them, and pinkish inverted 'Y' on front of head. Female caterpillar (80mm) larger than male one (50mm). **DISTRIBUTION** Eastern Australia, including south-eastern Qld, eastern NSW, Vic, Tas and south-eastern SA. **HABITAT AND HABITS** Found in wooded habitats. Caterpillars feed on wattles, particularly Cootamundra Wattle *Acacia baileyana* and Silver Wattle A. *dealbata*. Long hairs of caterpillar and its cocoon can easily puncture human skin and break off, causing mechanical skin irritation in some people.

Kristi Ellingsen

White Stemmed Wattle Moth ■ *Chelepteryx chalepteryx* WS up to 100mm

DESCRIPTION Adult moths brown with darker broad bands across wings. Pair of paler spots on each forewing, and reddish bases on hindwings. Male darker than female. Caterpillars reddish-brown, with conspicuous paler line along dorsal surface and numerous raised paler spots, each containing tuft of short bristles. **DISTRIBUTION** Eastern Australia, including Qld, NSW and Vic. **HABITAT AND HABITS** Found in forests, woodland and heaths, and adjacent urban areas. Caterpillars feed on leaves of several plant species, including wattles and lilies. When fully grown, at *c*. 70mm, caterpillar builds cocoon in tree crevice or under bark, which is covered with bristles. Bristles can cause pain and skin irritation.

Peter Street

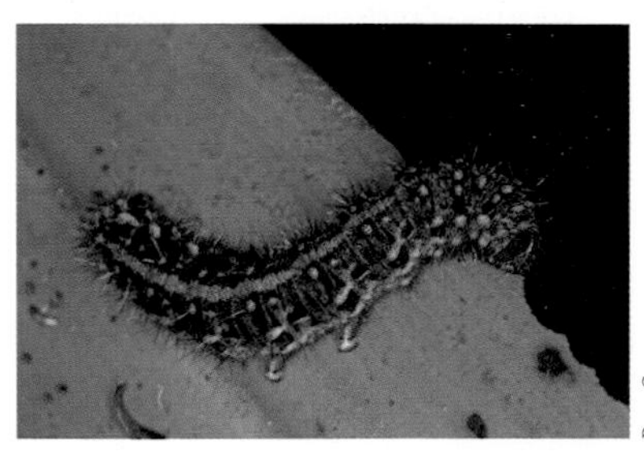

Peter Street

LEFT Adult; RIGHT Larva.

White Stemmed Gum Moth ■ *Chelepteryx collesi* WS 140–160mm

DESCRIPTION Adult moths greyish-brown with darker wavy lines, and bands of grey, yellowish and brown. Caterpillars strongly banded white and reddish-brown to blackish, with numerous raised yellow or reddish-brown spots, each with tufts of shorter reddish-brown or longer white bristles. **DISTRIBUTION** Much of south-eastern Australia, from southern Qld, through NSW and ACT, to Vic. **HABITAT AND HABITS** Found in forests, woodland, heaths and urban areas. Caterpillars known to feed on leaves of various eucalypts, angophoras and the Brush Box *Lophostemon confertus*. Once fully grown, at *c*. 120mm, caterpillar builds cocoon with 2 walls of silk, covered in numerous short, venomous bristles.

CSIRO ScienceImage (cc)

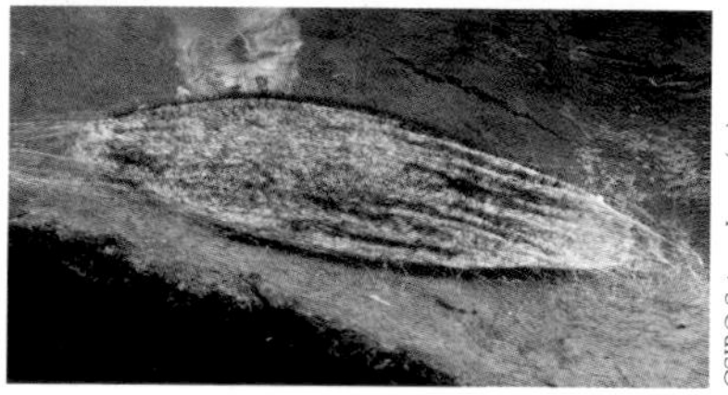

CSIRO ScienceImage (cc)

TOP Adult; LEFT Larva; RIGHT Cocoon.

Bucculatricidae (Scribbly Gum Moths)

Scribbly Gum Moth ■ *Ogmograptis scribula* WS 8mm

DESCRIPTION Generally undetectable as caterpillars, and most readily known by scribbled scars left on host tree when tree sheds its outer layer of bark. Adult moths greyish with darker brown blotches, and long, hair-like scales on trailing edges of wings. **DISTRIBUTION** Eastern Australia, including Qld, NSW and ACT. **HABITAT AND HABITS** Inhabits dry sclerophyll woodland. Eggs laid under new-forming bark of a eucalypt, including the Scribbly Gum *Eucalyptus rossii*. After hatching, tiny caterpillar burrows between old and new bark, forming characteristic scribble marks as it feeds. These widen as a caterpillar grows. Once fully developed caterpillar burrows to the surface, and pupates near base of tree or in leaf litter into the very small adult moth.

Peter Rowland/kapeimages.com.au

Marks left on tree by larvae.

Erebidae (Underwing, Tiger, Tussock Moths and Allies)

White Antennae Wasp Moth ■ *Amata nigriceps* WS 25–30mm

DESCRIPTION Adult moths black with orange translucent spots on each wing, black and orange-yellow banded abdomen, and white tips to black antennae. Male larger than female, but with thinner abdomen. Caterpillars orange and black, with longer white hairs.

Peter Rowland/kapeimages.com.au

DISTRIBUTION Coastal eastern Australia, from north-eastern Qld, to south-eastern NSW. **HABITAT AND HABITS** Found in open areas adjacent to sclerophyll forests and woodland, and regular visitor to urban parks and gardens, where female lays batches of small white eggs. Caterpillars and adults active by day. Hairs on caterpillar, which are also woven into its cocoon when pupating, can cause urticaria in sensitive people.

Perfect Tussock Moth ■ *Calliteara pura* WS 40–60mm

DESCRIPTION Moths off-white to greyish, covered with fine brownish wavy lines and flecks. Male more brownish on wings, yellowish on abdomen and has long, feather-like antennae. Caterpillars yellowish to orange and white, with numerous long hairs that change from yellowish to white after moulting. **DISTRIBUTION** From south-eastern Qld, through eastern NSW, to north-eastern Vic. **HABITAT AND HABITS** Common garden pest, but also found in woodland and other bushland. Female lays up to about 200 small, rounded eggs, and after hatching caterpillars feed on leaves of a variety of native and exotic plants, including roses, magnolias, Gymea Lily and Bird of Paradise. Once fully grown (at c. 40mm), caterpillar builds large cocoon of stinging hairs and moulted skins.

Nicholas Fisher

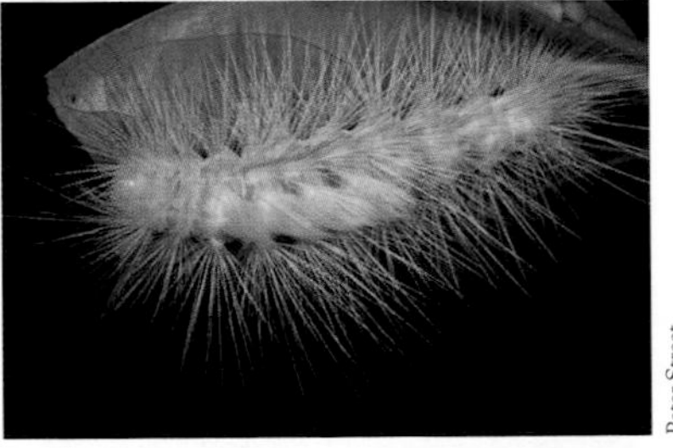

Peter Street

LEFT Adult; RIGHT Larva.

Vampire Moth

■ *Calyptra minuticornis* WS 45–50mm

DESCRIPTION Moths generally pale reddish-brown with silverish sheen, and leaf-like, with diagonal lines that mimic leaf veins. Caterpillars black with dark green stripes and orange head. **DISTRIBUTION** Eastern Qld and eastern NSW. More widely distributed in PNG, Indonesia, Malaysia, Taiwan and Japan. **HABITAT AND HABITS** Found mainly in rainforests, where caterpillar feeds on Snake Vine and other members of the moonseed family Menispermaceae. When fully grown (at *c.* 50mm) it pupates in the ground litter, in cocoon concealed between joined dead leaves. Moth normally feeds on fruit juices, but also feeds on blood of animals.

Dianne Clarke

Southern Old Lady Moth ■ *Dasypodia selenophora* WS up to 90mm

(Southern Moon Moth)

DESCRIPTION Adults brown above, patterned with darker lines and large blue, brown and black semicircular eye-spot on each forewing. Underneath orange-brown with small black spot. Caterpillars pale to dark brown, sparsely covered with short, stiff hairs, and with pale orange legs. **DISTRIBUTION** Mainly throughout eastern Australia, from northern Qld, to southern NSW. Some isolated records further inland, and more common in south. **HABITAT AND HABITS** Inhabits a range of habitats, where adults rest in plain sight, either on the ground or on trees or walls, with wings outspread, and frequently enter houses. Eggs laid in crevices, for example in cracks in tree bark. Caterpillars feed nocturnally on wattles, resting by day under leaves or branches.

Peter Rowland/kapeimages.com.au

Mistletoe Browntail Moth ■ *Euproctis edwardsii* WS 50mm

DESCRIPTION Adults orange-brown with pale spot on each forewing; abdomen darker with yellowish tufted tip. Caterpillars greyish-black, with paler white lines on dorsal surface, red spots along sides and long white hairs. **DISTRIBUTION** Throughout south-eastern Australia, except Tas. **HABITAT AND HABITS** Inhabits eucalypt woodland and adjacent areas, where caterpillars feed on mistletoe. Long, barbed hairs of caterpillars and their discarded skins break off with contact and are difficult to remove. They can cause urticarial wheals and papular eruptions that may persist for several days. They can also drift large distances in the wind and become troublesome to sensitive people over a wide area.

CSIRO Scienceimage (cc)

CSIRO Scienceimage (cc)

LEFT Adult; RIGHT Larva.

Reiner Richter

Yellow Tussock Moth

■ *Euproctis lutea* WS 30mm

DESCRIPTION Adult moths yellowish-brown and hairy with pair of paler zigzagging lines across forewings. Caterpillars black and hairy, with white spot on thorax and white longitudinal line along upper surface of abdomen. **DISTRIBUTION** Northern Australia, including WA, NT, Qld and NSW. **HABITAT AND HABITS** Found in a variety of wooded habitats, and urban parks and gardens. Feeds on leaves of freshwater mangroves, Cocky Apple and cultivated plants, including tomatoes and roses. Contact with hairs on caterpillar, egg, adult moth and even a site where caterpillars have been feeding can cause urticaria, which may become very itchy and swollen in sensitive individuals.

Black-bodied Browntail Moth ■ *Euproctis melanosoma* WS 30mm

DESCRIPTION Adult moths white with white abdomen, which becomes blackish with age, the female's with yellowish tip. Caterpillars black with 2 conspicuous red dots on dorsal surface towards rear end. **DISTRIBUTION** South-eastern Australia, including Qld, NSW, Vic and Tas. **HABITAT AND HABITS** Occurs in eucalypt woodland and adjacent areas. Caterpillars feed on a range of native and cultivated plants, including geraniums, roses, various grasses, Grape Vine *Vitis vinifera* and Brush Box *Lophostemon confertus*. When caterpillar is fully grown it pupates in a hairy cocoon on its food plant; adult moth emerges about 2 weeks later. The hairs cause irritation in humans.

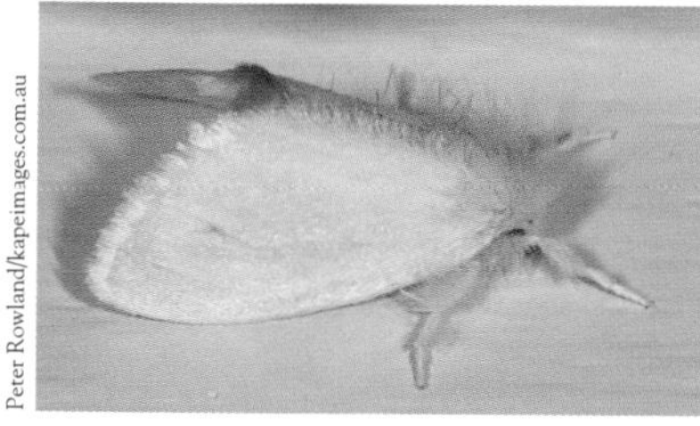
Peter Rowland/kapeimages.com.au

Kristi Ellingsen

LEFT Adult; RIGHT Larva.

Crimson Tiger Moth ■ *Spilosoma curvata* WS 30–35mm

DESCRIPTION Adult moths' wings creamish-yellow (forewings) to pinkish (hindwings), with variable pattern of broad black lines, blotches and spots. Hindwings entirely yellowish in some individuals, but always with black spot at centre of each. Thorax brown and abdomen crimson-pink, each segment with large black spot. Caterpillars dark, covered with brown hairs, becoming blackish towards black head, and with conspicuous longitudinal yellow line along upper surface. **DISTRIBUTION** Throughout much of eastern mainland Australia and Tas. **HABITAT AND HABITS** Occurs in a variety of habitats, mostly where low herbaceous plants grow. Female lays clutch of white, rounded eggs in a mass on a plant leaf, for example of dandelions, geraniums and beans. When fully grown (at c. 30mm), caterpillar pupates inside cocoon placed in crevice or on dry leaf.

Donald Hobern (cc)

Eupterotidae (Monkey Moths)

Lewin's Bag-shelter Moth ■ *Panacela lewinae* WS 30mm

DESCRIPTION Female moth dark brown with large tuft of hair on abdomen; male pale to reddish-brown with darker brown band on forewing and feathery antennae. Caterpillars brown with longitudinal zigzagging yellow line along sides and numerous long brown and white hairs. **DISTRIBUTION** Eastern Australia, from southern Qld, to southern NSW. **HABITAT AND HABITS** Found in a variety of wooded habitats. Caterpillars shelter together during the day in a protective shelter, normally a pair of leaves joined together with silk, emerging at night to feed on leaves of a variety of eucalypts, angophoras, box trees, Monterey Pine *Pinus radiata* and Native Cherry *Exocarpos cupressiformis*. When fully grown, caterpillar builds a cocoon in a curled leaf. Long hairs on caterpillar can cause skin irritation.

Donald Hobern (cc)

Geometridae (Geometer and Looper Moths)

Pink-bellied Moth ■ *Oenochroma vinaria* WS 60mm

DESCRIPTION Adults pinkish to brown, with dark-edged yellow line running across upper surface of wings, and with small, dark-edged, transparent spot on each forewing. Tips of forewing have blackish and grey, hooked tip. Underside of wings has large, dark purplish blotch. Closely resembles the **Mountain Vine Moth** *O. barcodificata* of NSW and Tas, which was separated from this species in 2009. Caterpillars typically brown with black back, variably spotted with white and with small, horn-like projections from spots. **DISTRIBUTION** Throughout most of Australia. **HABITAT AND HABITS** Found wherever larval food plants (mainly banksias, grevilleas and hakeas) grow, and readily enters gardens and buildings. When disturbed, caterpillar curls its head under its body, projecting its horns outwards.

Peter Rowland/kapeimages.com.au

Hesperiidae (Skippers)

Symmomus Ochre ■ *Trapezites symmomus* WS 42–50mm
(Symmomus Skipper, Splendid Ochre)

DESCRIPTION Black and pale-spotted skipper. Underside reddish-ochre with silver-white spots. Upperside dark brown suffused with orange-brown basally; forewing has large pale yellow spots; posterior median spot and hind-margin streak orange; hindwing has large median orange band. Underside reddish to yellow-ochre; forewing has coalesced yellow spots. Hindwing has large silver discal spot and almost continuous row of postmedian spots, becoming larger posteriorly. **DISTRIBUTION** Three subspecies: *T. s. sombra* Qld Wet Tropics; *T. s. symommus* central Qld to north-east Gippsland, Vic; *T. s. soma* south and south-west Vic. **HABITAT AND HABITS** Occurs in wetter forests, but is a regular visitor to gardens containing Matt Rush *Lomandra longifolia* (Xanthorreaceae). Flight a low, erratic, whirring-skipping near host plants. Settles regularly to feed or bask.

Rachel Whitlock

Limacodidae (Cup and Slug Moths)

Black Slug Moth ■ *Doratifera casta* WS 40mm

DESCRIPTION Adult moths have brown wings with 2 black dots on forewings. Caterpillars black, with white blotches and numerous off-white, fleshy spikes, 4 of which produce tufts of stinging hairs. Short legs. **DISTRIBUTION** Qld, NSW and Vic. **HABITAT AND HABITS** Found in wooded areas where its favoured food plants grow, mainly eucalypts, acacias and bottlebrushes. Female lays batches of about 40 brown, hairy eggs on the same leaf. After hatching, caterpillars initially feed together on the same leaf, but move to separate leaves as they grow in size. Stinging hairs of moths and caterpillars capable of causing minor to moderate skin irritation that can remain itchy for several days.

Andrew Brown (cc)

Painted Cup Moth ■ *Doratifera oxleyi* WS 50mm

DESCRIPTION Adult female moth mottled orange and brown; male has orange and white-tipped abdomen. Caterpillars pale greyish-green and white, with numerous fleshy spikes, and 4 pairs of tufted clear to yellowish stinging hairs. **DISTRIBUTION** South-eastern mainland Australia, from central eastern NSW, through southern Vic and Tas, to southern SA. **HABITAT AND HABITS** Found in eucalypt forests, woodland and adjacent urban parkland. Female lays batch of eggs on leaves of gum trees, covering cluster with stinging hairs from her own body. Stinging hairs of moths and caterpillars capable of causing an itchy rash that can persist for several days.

CSIRO ScienceImage (cc)

Lycaenidae (Blues, Coppers and Gossamer-winged Butterflies)

Plumbago Blue ■ *Leptotes plinius* WS 23–30mm

(Zebra Blue)

DESCRIPTION Small-tailed purple, black and white butterfly with strongly marked underside. Male upperside light purple with dark wing-margins. Underside densely spotted dark brown and white with submarginal spots. Tornus has black-and-green eye-spots and short, white-tipped filamentous tail. Female upperside dark, basally light metallic blue, with white postdiscal and postmedian spots; tornus has dark submarginal spots. Hindwing has white-tipped filamentous tail at tornus. **DISTRIBUTION** Torres Strait Islands and Qld Wet Tropics to Illawarra Coast and Cowra, NSW; also major inland centres including Alice Springs, NT. **HABITAT AND HABITS** Abundant anywhere the larval hosts are grown (ornamental Plumbago, *Plumbago auriculata*; Plumbaginaceae). Flight low, fluttering and erratic. Adults often settle to feed or bask. Males territorial. Larvae rarely ant attended.

Rachel Whitlock

Common Grass-blue ■ *Zizina otis* WS 20–23mm
(Grass Blue)

DESCRIPTION Tiny lilac and greyish-brown butterfly. Underside has indistinct postmedian and submarginal grey lunules. Male upperside bluish-lilac with darker margins. Underside pale brownish-grey with weakly defined postmedian and submarginal lunules. Female duller with upperside lilac dusting lighter and broader dark margins. **DISTRIBUTION** Two subspecies: *Z. o. labradus* nearly Australia-wide; *Z. o. labdalon* Torres Strait Islands and northern Cape York Peninsula, Qld. **HABITAT AND HABITS** One of Australia's most common and easily found butterflies, ubiquitous across most terrestrial habitats except rainforests and deserts. Flight low, brisk and fluttering, with adults settling often to bask or feed. Easily seen in urban areas, where it flies on nearly any lawn containing clovers *Trifolium* spp. (Fabaceae), which are the host plants. Larvae infrequently ant attended.

Peter Rowland/kapeimages.com.au

Nolidae (Tuft Moths)

Gum-leaf Skeletoniser ■ *Uraba lugens* WS 25–30mm

DESCRIPTION Adult moths greyish-white with darker wavy lines and flecks. Caterpillars brown with longitudinal lines of large and small yellow spots along back and sides, and numerous long white hairs. Head has long, ridged crown, formed from dried skins of past moults. **DISTRIBUTION** Throughout most of southern and eastern Australia, from north-eastern Qld, to south-western WA and Tas. **HABITAT AND HABITS** Occurs in forests and woodland. Female lays batches of eggs in long, parallel rows on leaves of trees such as eucalypts and Brush Box *Lophostemon confertus*. Young caterpillars feed together on leaves, forming long row that moves down leaf, leaving only leaf veins. With age, caterpillars feed more solitarily. Hairs of caterpillar can cause pain, redness and urticarial wheals.

Donald Hobern (cc)

Andrew Allen

LEFT Adult; RIGHT Larva.

Notodontidae (Bag-shelter Moths and Processionary Caterpillars)

Patterned Notodontid ■ *Aglaosoma variegata* WS 50–60mm

DESCRIPTION Caterpillars generally pale grey above, with yellowish-brown, raised lumps along dorsal surface and rows of elongated blue spots and reddish-brown mottling along sides. Numerous pale-tipped brown to black hairs, coronets of shorter stiffer hairs and hairy head. Adult moths have patterned black-and-white forewings with scattered small red spots, greyish hindwings and banded black-and-orange abdomen. **DISTRIBUTION** Eastern Australia from Atherton Tableland, Qld, through NSW, to Vic, possibly as far west as eastern SA. **HABITAT AND HABITS** Found in a range of habitats. Caterpillars feed on various plants, including wattles, banksias, she-oaks and rushes. Hairs of moths and caterpillars can cause severe skin rashes.

Dianne Clarke

Peter Street

LEFT Adult; RIGHT Larva.

Common Epicoma ■ *Epicoma melanosticta* WS 30mm

DESCRIPTION Adult moths have whitish forewings, each containing darker diagonal line and dark-edged golden spot, and dark brown hindwings edged with yellowish spots. Head and thorax tufted white, and abdomen blackish with golden spots and tip. Caterpillars blackish with long, brownish and grey hairs, and red band around rear of yellow-and-black head. **DISTRIBUTION** Throughout southern Australia, with records in each state and territory except NT. **HABITAT AND HABITS** Found in a variety of wooded habitats, where food plants grow. Female lays eggs in mass of hairs. After hatching caterpillars feed in groups on myrtles and tea trees, but feed individually when they are larger. Long hairs of moths and caterpillars capable of causing urticaria.

Donald Hobern (cc)

Bag-shelter Moth ■ *Ochrogaster lunifer* WS 40–50mm
(Processionary Caterpillar)

DESCRIPTION Adult moths grey to brown with variable conspicuous white patches and spots on upper surfaces of wings and darker band across undersurfaces, and long, tufted hairs on head, legs and body. Abdomen banded orange, black and yellow, with white tip. Caterpillars grey, with long white hairs and brownish head. **DISTRIBUTION** Suitable habitat throughout mainland Australia.

HABITAT AND HABITS Found in wooded habitats. Caterpillars known for forming long processionary train, as they walk head to tail in a line. They are voracious nocturnal feeders, mainly on wattles, and large groups are capable of stripping foliage completely from a shrub before moving to another. By day they huddle in a mass at the base of a food tree. Hairs of moths and caterpillars cause skin irritation.

Dianne Clarke

NYMPHALIDAE (NYMPHALIDS OR BRUSH-FOOTED BUTTERFLIES)

Red Lacewing ■ *Cethosia cydippe* WS 78–80mm

DESCRIPTION Medium-sized, bright red butterfly with broad black margins and white subapical patch. Underside dark and spotted with white marginal lunules. Male upperside bright red basally with broad dark margins and white marginal lunules. Forewing has white subapical patch; hindwing-margin strongly frilled. Underside reddish-brown with rows of white-edged black markings and submarginal spots with purplish borders. Forewing has orange basal area and white subapical patch. Females have broader black margins on upperside. **DISTRIBUTION** Torres Strait Islands, Cape York Peninsula and Qld Wet Tropics to near Townsville.

HABITAT AND HABITS Restricted to wet lowland rainforests; often along creeks and rivers and rather local. Flight a leisurely gliding with frequent flapping, a few metres above the ground, changing to strong wingbeats when alarmed. Both sexes settle often to bask. Larval hosts are native passion fruits (Passifloraceae); adults do not use plants grown in gardens for fruit.

John Nielsen

Peter Rowland/kapeimages.com.au

Peter Rowland/kapeimages.com.au

TOP *Upperwing*; BOTTOM *Underwing*.

Orange Lacewing

■ *Cethosia penthesilea* WS 65–70mm

DESCRIPTION Medium-sized orange butterfly with black margins and white subapical patch, underside densely spotted with white bands. Male upperside orange with dark median and discal markings and margins and white marginal lunules, forewing with margin enclosing white subapical band, hindwing with postmedian dark spots and strongly frilled margin. Underside orange with rows of black-fringed white bands and smaller black spots. Female similar. **DISTRIBUTION** NT Top End, including Melville Island and Groote Eylandt. **HABITAT AND HABITS** Coastal monsoon forests, often along waterways. Flight strong flutter with frequent gliding. Adults settle frequently to bask with their wings held at least partially open, gently moving the wings up and down. Larval hosts are native Passion Fruits (Passifloraceae).

Australian Rustic ■ *Cupha prosope* WS 47–53mm

(Bordered Rustic)

DESCRIPTION Medium-sized, orange-banded butterfly with black margins. Male upperside orange; light brown basally, with broad black margin. Forewing has margin enclosing small subapical orange spots; hindwing has obscured black submarginal spots. Underside pale ochre with narrow median margin and pale postmedian band. Forewing has subtornal black spot, and hindwing a submarginal row of purple-orange and black spots and pale marginal lunules. Female similar to male. **DISTRIBUTION** Torres Strait Islands and Cape York Peninsula, Qld, to near Grafton, NSW. **HABITAT AND HABITS** Found in rainforests and vine thickets. Flight a fairly quick flutter with occasional short bursts of planing. Adults usually found flying in clearings or along forest margins, settling often to bask or feed at flowers. Males territorial, perching within a few metres of the ground. Larval hosts include the Flintwood *Scolopia braunii* and Governer's Cherry *Flacourtia indica* (both Salicaceae).

Dianne Clarke

Black and White Tiger

■ *Danaus affinis* WS 62–65mm
(Swamp Tiger)

Thomas Rowland/kapeimages.com.au

DESCRIPTION Medium-sized, black-and-white banded butterfly. Underside has postmedian pale orange markings. Male upperside dark with median to basal white areas; tiny postmedian and submarginal white spots. Forewing has broad white subapical band; underside similar but paler with orange postmedian markings in hindwing. **DISTRIBUTION** *D. a. affinis* in north-west WA, Top End, NT and Cape York Peninsula, to Port Macquarie area, NSW. **HABITAT AND HABITS** Restricted to open paperbark wetlands and margins of estuarine waterways; populations rather local. Flight a leisurely flutter with frequent gliding, always within a few metres of the ground. Not migratory but individual adults can occasionally be found flying through urban or subcoastal areas some distance from breeding grounds. Individuals congregate near the larval hostplant, the climbing Mangrove Milkweed *Cynanchum carnosum* (Apocynaceae), and both sexes patrol the habitat searching for mates.

Wanderer ■ *Danaus plexippus* WS 92–93mm

(Monarch Butterfly)

Peter Rowland/kapeimages.com.au

Peter Rowland/kapeimages.com.au

TOP Adult; BOTTOM Larva.

DESCRIPTION Large, tawny-orange butterfly with long, acute forewings, bold black veins and margins with double row of small white submarginal spots. Male upperside tawny-orange with slightly paler subapical and postdiscal spots, veins mostly delineated in black, and black margin enclosing double row of white spots. Underside paler; hindwing creamy-orange. Female has broader black markings on veins; sex brand absent on hindwing. **DISTRIBUTION** Torres Strait Islands, Cape York Peninsula, Qld, to Adelaide and Port Lincoln, SA; also Perth area, WA, and scattered records across NT. **HABITAT AND HABITS** Widespread in disturbed urban areas, farmland and open grassland. Flight powerful and direct; often some gliding. Typically flies within 3m of the ground but may go much higher. Native to North America, this species island hopped to Australia during the 1800s following introduction of its host plants with European colonization of Australia. The boldly striped larvae feed on introduced milkweeds (Apocynaceae) including the Swan Plant or Wild Cotton *Gomphocrapus fruticosa* and Red Milkweed *Asclepias curassavica*.

Peter Street

Common Crow Butterfly

■ *Euploea corinna* WS 69–72mm

DESCRIPTION Medium-sized, black-and-white marked butterfly with broad wings. Male upperside dark with white spots. Forewing has large postmedian spots, smaller subapical and submarginal spots, and median linear grey sex brand. Hindwing has submarginal band of narrow white spots and very small marginal spots. Underside similar but greyish. Sex brand absent in female. **DISTRIBUTION** North-west WA, Top End, NT and Cape York Peninsula, Qld, to north-east NSW; adults occasionally reach to ACT and eastern Vic. **HABITAT AND HABITS** Found in coastal open forests and rainforest margins but well adapted to gardens. Flight strong with frequent glides. Settles regularly to feed at flowers or dried leaves of certain plants to obtain chemicals for pheromone production. Larval hosts include the ornamental Oleander *Nerium oleander* and milkweeds (all Apocynaceae).

Common Eggfly ■ *Hypolimnas bolina* WS 72–74mm

(Varied Eggfly)

DESCRIPTION Large, black-and-white butterfly with purple iridescence. Female has orange forewing-patch. Male upperside black with median white patches ringed with shining iridescent purple and smaller subapical white patch. Underside pale brown with diffuse white median bands, submarginal white spots, and submarginal and marginal lunules. Female has broader white median bands, often suffused with blue iridescence, and forewing with orange subtornal patch; very variable. **DISTRIBUTION** North-west WA, Top End, NT and Cape York Peninsula, Qld, to north-east NSW; vagrants recorded as far south as ACT, Vic, Tas and SA. **HABITAT AND HABITS** Occurs in gallery and open forests; well adapted to urban areas. Flight a strong, erratic flutter with some gliding. Adults perch in sunny patches with wings partly open. Males very territorial. Females fly low, looking for host plants, which include the Pastel Flower *Pseuderanthemum variable*, Ganges Bluebell *Asystasia gangentica* (both Acanthaceae), and *Sida retusa* (*Sida rhombifolia*; Malvaceae).

Rachel Whitlock

Sharon McGrigor

TOP Male; BOTTOM Female.

Chocolate Argus ■ *Junonia hedonia* WS 50–51mm

DESCRIPTION Medium-sized, light or reddish-brown butterfly with somewhat leaf-shaped wings. Male upperside reddish-brown with darker discal, median and submarginal bands and postmedian row of dull red eye-spots. Underside darker with diffuse blue-grey postmedian band, creating a dead-leaf illusion. Hindwing has postmedian row of cream spots. Female similar; mostly paler. **DISTRIBUTION** Top End, NT and Qld from Torres Strait Islands and Cape York Peninsula, to south-east Qld. **HABITAT AND HABITS** Found at margins of freshwater wetlands, including *Melaleuca* forests; always local. Flight a strong flutter with some jerky, flat-winged planing, fairly low over understorey or emergent vegetation. Often settles to bask or feed at flowers. Larval host is the Hygro *Hygrophila angustata*, an emergent herb. The Northern Argus *J. erigone* is a vagrant on NT islands.

Thomas Rowland/kapeimages.com.au

Meadow Argus ■ *Junonia villida* WS 40–43mm

DESCRIPTION Small brown butterfly with large orange eye-spots and discal bands. Underside pale buff with subtornal forewing eye-spot. Male upperside brown. Forewing has very large black and blue/white-centred orange subtornal eye-spot, orange discal bands and smaller subapical eye-spot. Hindwing has large subapical and smaller subtornal orange eye-spots, and scalloped, alternately pale and dark margins. Underside buff, basally orange on forewing; postmedian area has small eye-spots or diffuse darker band; forewing has large subtornal eye-spot. Female larger, with more rounded wings. **DISTRIBUTION** Australia-wide except west-coast Tas. **HABITATS AND HABITS** Adults breed in savannah or dry grassland, but have adapted well to disturbed urban areas; migratory. Flight low and rapid, with regular long planing. Often settles on the ground or on low flowers to bask or feed. Familiar sight in gardens, parkland and even cities. One of Australia's most abundant species. Larval hosts diverse, but include weedy plantains *Plantago* spp. (Plantaginaceae)

Peter Rowland/kapeimages.com.au

Donald Hobern (cc)

Donald Hobern (cc)

TOP Upperwing pattern; BOTTOM Underwing pattern.

Orange Bush-brown

■ *Mydosama terminus* WS 35–38mm

DESCRIPTION Small, tawny-orange and black butterfly with eye-spots. Upperside tawny-orange with dark margins. Forewing orange-brown medially with paler postdiscal patch and black eye-spot. Apex, costa and termen broadly black. Hindwing reddish-brown with three postmedian eye-spots and two dark submarginal bands. Underside pale ochre with dark median line, paler postmedian line, paired wavy brown submarginal lines, and row of large, well-defined, postmedian eye-spots. Female larger. **VARIATION** Dry-season form paler with underside eye-spots, small or absent. **DISTRIBUTION** Torres Strait Islands and Cape York Peninsula to near Childers, Qld. **HABITAT AND HABITS** Restricted to coastal open forests and adjacent rainforest margins, often along rivers or creeks and sometimes in rainforest understorey. Flight low, leisurely bobbing. Often settles to bask. Larval hosts are grasses (Poaceae).

Blue Tiger ■ *Tirumala hamata* WS 70–72mm

DESCRIPTION Medium-sized, black-and-blue spotted butterfly. Male upperside dark with numerous light blue spots and streaks. Hindwing has obscure grey sex brand opening to small, conspicuous pouch on underside; underside paler, suffused greyish. Female similar but lacks sex brand and underside pouch. **DISTRIBUTION** Extreme north-west WA, NT Top End, Torres Strait Islands and Cape York Peninsula, to north-east NSW; migrants reach ACT and north-east Vic. **HABITAT AND HABITS** Breeds in vine thickets and coastal forests. Populations often swell after heavy rainfall, resulting in large-scale migrations in coastal areas. Millions of butterflies may pass through an area over a few weeks, making a spectacular sight. Both sexes readily settle to feed at flowers and congregate overnight to roost, festooning trees. Flight a strong, steady flutter with some gliding. Larval hosts are climbing milkweeds such as *Cynanchum carnosum* and *Secamone elliptica* (Apocynaceae).

Scott Eipper

'Orange' Sword-Grass Browns ■ *Tisiphone abeona* spp. WS 47–55 mm

DESCRIPTION Medium-sized, dark brown butterflies with broad orange forewing-patch. Male upperside dark. Forewing has subapical, discal and postmedian orange patches and large, dark postmedian eye-spot. Hindwing has obscured apical eye-spot and red and black eye-spot at tornus. Underside forewing similar but paler. Hindwing has cream postmedian line and two submarginal lines. Female has slightly broader markings and is larger. **VARIATION** Three subspecies: *T. a. abeona* as described; *T. a. aurelia* has subtornal orange-and-black eye-spot; *T. a. albifascia* has additional thin submedian white lines on undersides of both wings. **DISTRIBUTION** *T. a. aurelia*: Port Macquarie area to near Newcastle, NSW; *T. a. abeona*: Newcastle area, NSW, to Bateman's Bay area, NSW; *T. a. albifascia*: Tallaganda Range, NSW, to extreme south-east SA. **HABITAT AND HABITS** Found in tall, open forests from sea level to more than 1,000m altitude. Flight a low, leisurely bobbing flutter in the understorey and along forest margins. Males territorial. Larval host plants are sword grasses *Gahnia* spp. (Cyperaceae).

Donald Hobern (cc)

Cruiser ■ *Vindula arsinoe* WS 75–82mm

DESCRIPTION Large, orange, or brown, white and orange, butterfly with black spots. Hindwing has conspicuous eye-spots. Male orange with row of postmedian spots and submarginal band. Forewing has dark discal and postdiscal streaks; hindwing has two orange and black postmedian eye-spots. Underside paler, with postmedian and submarginal markings diffuse. Hindwing has additional narrow, dark median band. Female darker, with similar patterns, but with additional postmedian white band; hindwing has band suffused with orange posteriorly; underside has pale postmedian and submarginal bands. **DISTRIBUTION** Qld, from Torres Strait Islands, Cape York Peninsula and Wet Tropics, to near Mackay. **HABITAT AND HABITS** Found along margins of rainforests, gallery forests and vine thickets. Flight rapid with much gliding; males territorial, perching several metres above the ground. Females elusive. Larval hosts are various passionfruits (Passifloraceae).

Peter Rowland/kapeimages.com.au

Peter Rowland/kapeimages.com.au

TOP Female; BOTTOM Male.

PAPILIONIDAE (SWALLOWTAIL BUTTERFLIES)

Peter Rowland/kapeimages.com.au

Blue Triangle ■ *Graphium choredon* WS 53–60mm

DESCRIPTION Medium-sized, blue-and-black butterfly with angular wings. Upperside black with median row of translucent blue spots broadly joined to form a band; hindwing-band has additional row of submarginal lunules. Underside paler, grey-brown with blue spots and bands broader with diffuse margins. Hindwing has median row of red-and-black spots. **DISTRIBUTION** Torres Strait Islands and Cape York Peninsula, Qld, south to NSW–Vic border; rare vagrant or introduction to ACT. **HABITAT AND HABITS** Occurs in rainforests and gardens. Common wherever the Camphor Laurel *Cinnamomum camphora* (Lauraceae) grows. Adults easily seen along forest margins and in urban gardens; flight rapid and dashing, but not frenetic. Courtship spectacular, sometimes with several males hovering above a female, swooping down to mob her with their hindwing feather-like scent-tufts everted. Occasionally confused with the Ulysses Butterfly (opposite), which has iridescent rather than translucent blue wings.

Sharon McGrigor

Sharon McGrigor

TOP Female upperwing pattern; BOTTOM Male underwing pattern.

Cairns Birdwing

■ *Ornithoptera euphorion* WS 125–180mm

DESCRIPTION Very large butterfly, iridescent green and black, or black, grey and yellow. Male forewing-upperside velvet-black with iridescent green margins. Hindwing iridescent green with black submarginal spots and margin. Forewing underside black with restricted iridescent green discal, postmedian and submarginal spots. Hindwing similar to upperside but paler, with gold submarginal areas. Female upperside black with restricted grey-white discal, postmedian and submarginal spots. Hindwing has submarginal spots enclosed by dirty-yellow and white lunules; underside similar but brighter. **DISTRIBUTION** Qld Wet Tropics coast and tablelands, from Cooktown south to near Mackay. **HABITAT AND HABITS** In wet rainforests, breeds in suburbia if the larval hosts *Aristolochia* spp. (Aristolochiceae) are grown in gardens. Flight strong and soaring in the canopy, descending in the cooler hours of the morning and evening to feed at flowers. Both sexes rest during the heat of the day. Males territorial.

Orchard Butterfly ■ *Papilio (Princeps) aegeus* WS 102–108mm
(Orchard Swallowtail)

DESCRIPTION Large, black-and-white butterfly with broad wings and red hindwing-spots. Male upperside black. Forewing has subapical cream spots; hindwing has median cream band and red tornal spot. Hindwing underside has median and submarginal rows of grey (inner), blue (middle) and red (outer) spots. Female forewing black with outer half grey-white; veins black. Hindwing has white median band and submarginal rows of blue (inner) and red (outer) lunules; underside paler. **DISTRIBUTION** Originally found in Cape York Peninsula, Qld, south through NSW, ACT and Vic, to south-east SA, now recorded from all mainland capitals and Alice Springs, NT. **HABITAT AND HABITS** Occurs in wetter coastal forests but adapted to suburbia and now occurs anywhere *Citrus* spp. (Rutaceae) grow. Flight a strong but leisurely jinking, becoming faster and exaggerated if disturbed. Both sexes often rest on foliage with wings held flat and feed at flowers with wings fluttering.

Sharon McGrigor

TOP Female; BOTTOM Male.

Ulysses Butterfly
■ *Papilio (Princeps) ulysses* WS 100–108mm
(Ulysses Swallowtail)

DESCRIPTION Large, iridescent blue butterfly; each hindwing has spatulate tail. Upperside iridescent blue with intense black margins. Male forewing has submarginal row of broad, lanceolate grey sex brands. Female upperside has blue areas slightly less extensive than in male; hindwing has submarginal blue lunules. Underside brown-black. Forewing has broad, diffuse greyish band; hindwing has large brown submarginal lunules and orange spot at tornus. **DISTRIBUTION** Qld, from Torres Strait Islands, Cape York Peninsula and Wet Tropics coast, south to near Mackay. **HABITAT AND HABITS** In rainforests, often seen as flashes of blue from some distance away. Flight frenetic, vigorous and jinking. Males patrol forest margins or creeks; females elusive. Red flowers are favoured nectar sources. Adults skittish and keep wings in motion, making photography difficult. Larval hosts are evodias *Melicope* spp. (Rutaceae).

Steve and Alison Pearson

Peter Rowland/kapeimages.com.au

TOP Upperwing pattern; BOTTOM Underwing pattern.

Pieridae (Pierid Butterflies, Whites and Yellows)

Northern Jezebel ■ *Delias argenthona* WS 62mm
(Scarlet Jezebel)

Dianne Clarke

DESCRIPTION Black-and-white butterfly. Male upperside creamy-white. Forewing has black apex enclosing narrow white subapical spots. Hindwing has black margin becoming grey postmedially, more or less enclosing submarginal spots; underside pattern faintly visible. Forewing underside similar but with postdiscal white spot. Hindwing underside basally yellow with outer half black and enclosing red postdiscal and submarginal spots edged with white. Female darker, with all markings larger. **DISTRIBUTION** Two subspecies: *D. a. fragelacta* north-western WA and NT Top End; *D. a. argenthona* Torres Strait Islands and Cape York Peninsula, Qld, to central NSW coast; vagrant to southern NSW, ACT and northern Vic. **HABITAT AND HABITS** Occurs in coastal forests, dispersing into adjacent areas including gardens. Winter or montane species in North Queensland. Flight a slow flutter with regular soaring several metres above the ground. Larval host plants are mistletoes (Loranthaceae).

Cabbage White ■ *Pieris rapae* WS 40mm
(Small Cabbage White; Cabbage Moth)

DESCRIPTION Small white butterfly with black forewing-spots. Underside of hindwing creamy-yellow. Male upperside chalky-white; forewing apex and postmedian spot black; hindwing has one black apical spot. Forewing underside white with two postmedian black spots; apex and hindwing creamy-yellow with dark suffusion. Female darker with two postmedian black spots on forewing upperside. **DISTRIBUTION** Much of south-east and south-west Australia, with isolated records from central Australia and Qld Wet Tropics. **HABITAT AND HABITS** Introduced and familiar species that occurs anywhere cultivated or weedy brassicas (Brasssicaceae) grow. In gardens, larvae are major pests of cabbage, broccoli and kale; weedy mustards and nasturtiums are also used. Flight a low, busy flutter, with adults settling regularly to bask or feed at flowers.

Peter Rowland/kapeimages.com.au

Peter Rowland/kapeimages.com.au

LEFT Male; RIGHT Female.

Psychidae (Bagworm Moths)

Saunders' Case Moth ■ *Metura elongatus* WS 30mm (male)

DESCRIPTION Male moth's wings brown with paler veins; head and thorax have orange specks; abdomen banded dark brown and orange. Female moth wingless. Caterpillars bright black and orange on head and first body segments; remaining body (inside case) whitish. **DISTRIBUTION** Eastern Australia in Qld, NSW, Vic and Tas. **HABITAT AND HABITS** Occurs in various habitats, including forests, woodland and gardens. Caterpillars feed on a wide variety of plants, including eucalypts, paperbarks and tea trees. Caterpillar houses itself in a large, soft case, with pieces of twigs attached longitudinally by a silken web. Once fully grown, it attaches the case to a vertical surface and pupates inside. After emerging from pupa male moth leaves pupal case protruding, while female remains in her case. Following mating, female lays eggs within the case, after which she dies.

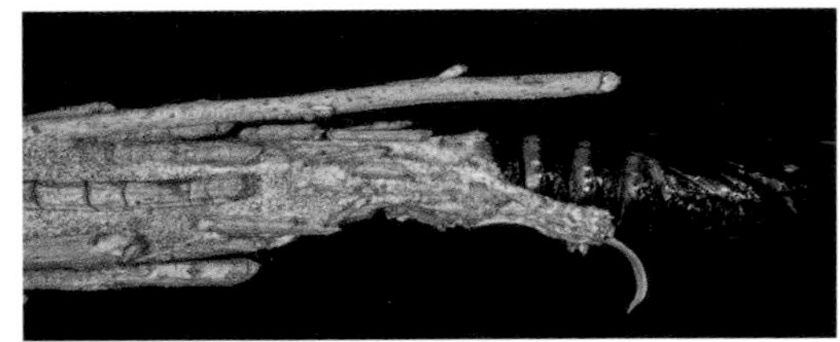
Peter Rowland/kapeimages.com.au

Case Moth ■ *Conoeca guildingi* WS 20–25mm

DESCRIPTION Female moth grey with darker speckling; hindwings have dark margins; abdomen banded with yellow and black. Male's forewings whitish with darker speckles; hindwings blackish; thorax yellowish and abdomen blackish. Caterpillars patterned brown and whitish on head and front part of body. **DISTRIBUTION** Eastern Australia in Qld, NSW, Vic and Tas. **HABITAT AND HABITS** Occurs in various wooded areas, including gardens. Larvae live in tapering case of finely chewed bark, to 30mm long, and have been found feeding on myrtles (Myrtaceae) such as the Woolly Tea Tree *Leptospermum lanigerum* and kunzeas, as well as rushes (Juncaceae) and peas (Fabaceae).

Peter Rowland/kapeimages.com.au

Apidae (Bees)

European Honeybee ■ *Apis (Apis) mellifera* TL 16mm
(Common Honeybee, Western Honeybee)

DESCRIPTION Generally yellowish-brown all over, with blackish-brown bands on abdomen and orange-brown wash. Some variation in colour and pattern between individuals.

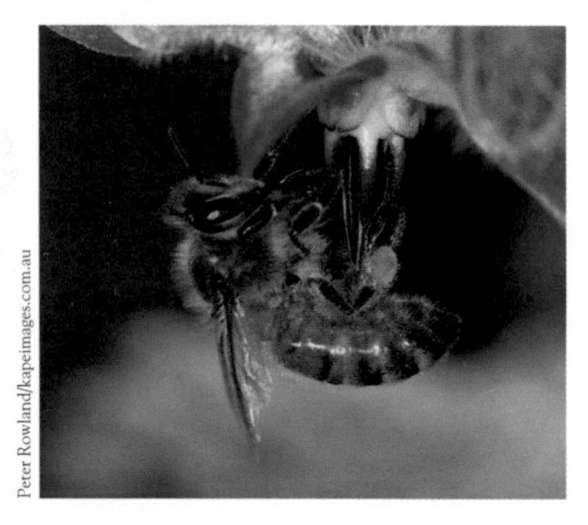
Peter Rowland/kapeimages.com.au

DISTRIBUTION Introduced into Australia by early European settlers, and now widespread. **HABITAT AND HABITS** Occurs in forests, woodland, heaths and urban areas, wherever nectar-producing flowering plants are found, living in large communal nest (hive). The sting, which it uses to protect the colony, is barbed and stays embedded in its victim when it flies away (the bee dying as a result), while the venom gland continues to pump the venom into its victim. On average about 1,000 people are hospitalized due to allergic reactions (anaphylaxis) to bee venom each year in Australia.

Teddy-bear Bee ■ *Amegilla (Asaropoda) bombiformis* TL 15mm

DESCRIPTION Stocky, golden-yellow bee, with 6 (female) or 7 (male) variable-width, blackish, hairless cross-bands on abdomen. Remainder of body covered in short hairs, giving it a 'furry' appearance. **DISTRIBUTION** Mainly south-eastern Australia, from south-eastern Qld, to southern Vic. **HABITAT AND HABITS** Occurs in various habitats where flowering plants grow. Flight rapid and jerky, often hovering periodically and settling briefly on flowers to collect pollen; wings make a low-pitched humming sound as it flies. Solitary bee, although several may occupy the same area, excavating a nest in the ground (typically in an earthen bank) that contains several watertight cells. Single egg is laid in each cell, which is provided with a supply of pollen to feed the larva after hatching, before cell is sealed to protect it.

Jenny Thynne

Blue-banded bees ■ *Amegilla* spp. TL 15mm

DESCRIPTION Adults have black abdomen with bright metallic blue hairy bands; remainder of body covered in gold-and-white, furry hair, including tops of legs. Eyes large and greenish-yellow. Male distinguished from female by extra abdominal band and lack of black on clypeus (part of face near mandibles). Wings smokey-brown and short. **DISTRIBUTION** Tropical and subtropical regions of Australia. Reported from parts of South-east Asia. **HABITAT AND HABITS** Lives in forests, woodland and heaths, including urban areas. Females inhabit burrows in soil, typically in banks of rivers or streams, or in loose joints in brick walls, and often in close association with others. Obtains pollen from flowers by 'buzz' pollination, using sonic vibrations to release pollen. Capable of stinging but less aggressive than other bees.

Peter Rowland/kapeimages.com.au

Stingless Bee ■ *Tetragonula carbonaria* TL 3–4mm

(Sugarbag Bee)

DESCRIPTION Small, stingless bee. Adults blackish with brownish legs and transparent, simply veined wings. **DISTRIBUTION** Endemic to coastal and near-coastal eastern Australia, mainly north of Sydney. **HABITAT AND HABITS** Occurs in open forests and woodland. Nests formed in the ground and bases of trees, the entrance lined with a mixture of beeswax and resin (cerumen). Forms spiral-shaped honeycomb structures in nest to house eggs and produces sugarbag honey, traditionally sought after and eaten by indigenous Aboriginal people, but also produced commercially in small quantities. Workers may travel long distances to collect pollen from various plants including cycads; also known to pollinate orchids.

Dianne Clarke

Kristi Ellingsen

LEFT Foraging adult; RIGHT Nest.

Chequered Cuckoo Bee ■ *Thyreus caeruleopunctatus* TL 11–15mm

DESCRIPTION Striking, stocky black-and-blue bee. Mainly black, with numerous dense, hairy blue spots and patches, including 4 on upper surface of each abdominal segment, 2 on front of thorax (scutum). Face mostly blue with large black eyes; wings brownish-black with iridescent sheen. **DISTRIBUTION** Throughout northern and eastern Australia. **HABITAT AND HABITS** Inhabits forests, woodland and heaths, including urban areas. Parasitizes blue-banded bees of the genus *Amegilla* (see p. 143), although exact species targeted is not adequately recorded. Female lays egg in unguarded egg chamber of its host, and after hatching larva eats pollen provision intended for the host larva, leaving no food for it and thus starving it. Egg or larva of host may also be eaten. Adult has thickened scutellum on thorax for protection against stings from host bee.

Dianne Clarke

Neon Cuckoo Bee ■ *Thyreus nitidulus* TL 10–15mm

DESCRIPTION Stocky, striking iridescent black-and-blue bee. Predominantly black with incomplete dense, hairy blue bands on abdomen, 4 blue spots on top of thorax, mostly blue face, and brownish-black wings with purple sheen. **DISTRIBUTION** Broad coastal band through northern and eastern Australia. **HABITAT AND HABITS** Inhabits forests, woodland and heaths, including urban areas, and parasitizes other native bees. Female parasitizes the **Blue-banded Bee** *Amegilla cingulata*, seeking out its nest and sneaking in when unguarded to lay a single egg in an unsealed, already occupied cell. After hatching, cuckoo bee larva eats pollen provision intended for *Amegilla* larva, leaving no food and thus starving host larva. Eggs and larva of host species may also be eaten. Adult has thickened scutellum on thorax for protection against stings from host bee.

Dianne Clarke

Peacock Carpenter Bee ■ *Xylocopa (Lestis) bombylans* TL 16–18mm

DESCRIPTION Metallic purplish to yellowish-green with blackish wings. **DISTRIBUTION** Eastern Australia from Sydney, NSW, to northern Cape York Peninsula, Qld. **HABITAT AND HABITS** Occurs in wooded areas, including forests, woodland, heaths and gardens. Emits a deep, resonating buzz as it moves between flowers to feed on nectar and pollen. Nests excavated in wood, and consist of a long tunnel (*c*. 30cm) with a series of brood chambers, each chamber containing a single egg and a ball of pollen for larva to feed on and sealed off for protection.

Jenny Thynne

Chrysididae (Cuckoo Wasps)

Cuckoo wasps ■ *Chrysis* spp. TL 10mm

(Emerald wasps)

DESCRIPTION Metallic blue-green to purplish. Body heavily pitted and rear of thorax with median projection. **DISTRIBUTION** About 25 species widely distributed in eastern Australia. **HABITAT AND HABITS** Found in a variety of vegetated habitats and often seen around external walls of buildings or around windows. Female lays eggs in nest of mud-nesting wasp species, where young hatch and feed on the paralysed larvae left for the young of the mud-nesting wasp or on the mud-nesting larvae themselves. If threatened, adults curl up to expose hardened upper body and protect their more vulnerable underside.

Kristi Ellingsen

Cuckoo Wasp ■ *Stilbum cyanurum* TL 20mm

DESCRIPTION Colour variable, but typically metallic blue-green to purplish in Australia. Northern hemisphere individuals more golden or reddish. Body heavily pitted, rear of thorax with median projection and sides of thorax with 4 downwards-pointing teeth. Size varies greatly; largest of the cuckoo wasp species. **DISTRIBUTION** Widely distributed in eastern hemisphere and throughout Australia. **HABITAT AND HABITS** Found in a variety of vegetated habitats. Female lays eggs mainly within nest of mud-nesting wasp species, where young hatch and feed on paralysed larvae left for the young of the mud-nesting wasp. Some megachilid bees are also used as hosts. If threatened, adults curl up to expose hardened upper body and protect their more vulnerable underside.

Bernhard Jacobi

Crabronidae (Sand Wasps)

Cicada-killer Wasp ■ *Exeirus lateritius* TL 40mm

DESCRIPTION Dark brown and golden-yellow, with banded abdomen, yellow legs, head and antennae, and yellowish-brown, membranous wings. Eyes large and black. **DISTRIBUTION** Throughout Australia. **HABITAT AND HABITS** Found in forests, woodland, heaths and urban areas, in warm, dry regions. Solitary hunter of cicadas. Once located, a cicada is stung and paralysed by the wasp, making it drop to the ground. Wasp then drags the cicada to a multicellular underground nest, up to 60cm underground, and lays a single egg upon it. Wasp seals the nest chamber and, after hatching, larva feeds on body of cicada. Males patrol open areas in search of hunting females to breed with.

Noel Karlson

Formicidae (Ants)

Golden-bearded Sugar Ant ■ *Camponotus aurocinctus* TL 8–14mm

DESCRIPTION Generally dark red to blackish, the abdomen (gaster) usually with variable golden bands, from single band (tergite) to being fully covered. Legs lack erect hairs. Workers have cluster of up to 30 elongated curved hairs on underside of head near base of mouthparts (labia). Major workers have larger head than minor workers. **DISTRIBUTION** Endemic to Australia, and found from western Qld, western NSW and north-western Vic, through SA and southern NT, to central WA. **HABITAT AND HABITS** Shows marked preference for areas of sandy soils, where it nests underground in colonies. Workers forage on the ground and in low vegetation for a variety of plant material, including seeds, flowers, nectar and honeydew, but also collect small arthropods. Ants of this genus do not have a sting, but are able to spray offensive chemicals from the abdomen.

Thomas Rowland/kapeimages.com.au

Honey Ant ■ *Camponotus inflatus* TL 8–16mm
(Honeypot Ant)

DESCRIPTION Generally black with yellow rings around abdomen. Body covered with sparse hairs. Major workers (repletes) often have distended golden-brown abdomens, with broad black bands. **DISTRIBUTION** Central Australia, including NT and SA. **HABITAT AND HABITS** Occurs in mulga woodland. Minor workers feed on honeydew secreted by the **Red Mulga Lerp** *Austrotachardia acaciae*, or on nectar secreted by phyllodes of mulga tree. Nests excavated in sandy soil below mulga trees, up to 2m deep, typically on shaded side of tree to reduce effects of hot sun. Major workers remain in nest and are fed honey by minor workers, leading to their abdomens becoming swollen with honey as storage to feed colony during times of drought. Popular bush tucker for indigenous Australians, who dig down to find swollen ants and suck out stored honey.

Greg Hume (cc)

Black-headed Sugar Ant ■ *Camponotus nigriceps* TL 8–14mm

DESCRIPTION Head and abdomen black, remainder of body brownish. Body smooth with sparse short hairs, longer and most numerous on abdomen. Major workers (soldiers) larger than minor workers, and with larger head and more powerful jaws. **DISTRIBUTION** Throughout eastern and southern Australia, from north-eastern Qld, through NSW, Vic and SA, to southern WA. **HABITAT AND HABITS** Inhabits open grassland and dry sclerophyll forests. Forages mainly at night for nectar honeydew from insects that secrete it, but also feeds on small invertebrates. Often enters homes, attracted by sugary substances. Nest excavated in sandy soil in open ground, or occasionally under a rock, and contains a single queen. Ants of this genus do not have stings, but aggressively defend their nest and are able to inflict a painful bite and spray offensive chemicals from the abdomen.

Lou Wolfers

Meat Ant ■ *Iridomyrmex purpureus* TL 8–12mm

DESCRIPTION Long-legged, with large, reddish-brown head and dark metallic bluish-black abdomen. **DISTRIBUTION** Throughout mainland Australia, and overlaps with other similar species also known as meat ants. **HABITAT AND HABITS** Inhabits woodland, grassland and adjacent urban areas with sand or gravel. Forages by day on a variety of plant and animal matter, both dead and living. Can form symbiotic relationships with caterpillars and scale insects that exude sugary substances, on which it feeds, providing protection from intruders in return. Constructs large underground nests that can house more than 60,000 individuals; several adjacent nests may form part of a single colony. Very aggressive towards intruders, readily biting repeatedly with powerful jaws, and spraying a pungent liquid from the anal gland.

Peter Rowland/kapeimages.com.au

Giant Bull Ant ■ *Myrmecia brevinoda* TL 26mm

(Bulldog Ant)

DESCRIPTION Variable, but generally reddish-brown, with black abdomen, large eyes and powerful, forwards-projecting jaws. Typically has a long body, and is among the longest ants in the world. **DISTRIBUTION** Eastern Qld, NSW and eastern Vic. **HABITAT AND HABITS** Found in rainforests, open forests and woodland. Typically nests in soil or under logs and rocks. Abdomen has a stinger that is used to envenom and subdue insect prey, but also feeds on honeydew and other sweet secretions from plants and other insects. Sting painful, and venom has caused severe allergic reactions and fatalities in humans.

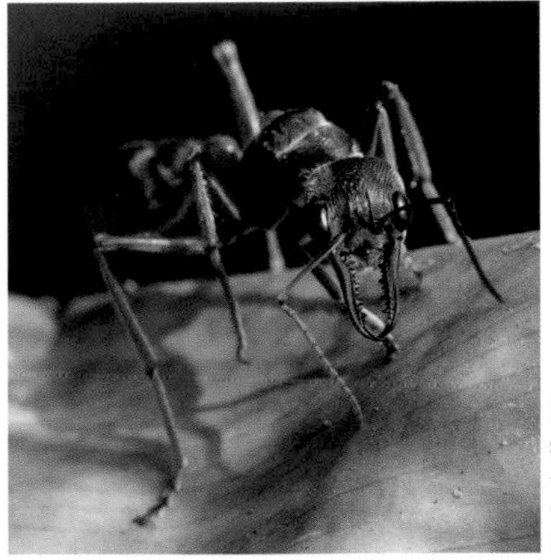

Peter Rowland/kapeimages.com.au

Jumper Ant ■ *Myrmecia pilosula* TL 17mm

(Jack Jumper)

DESCRIPTION Mainly black and orange to reddish-brown, with large, yellowish, forwards-pointing jaws and large eyes. **DISTRIBUTION** Coastal eastern Australia, from northern Qld to eastern Vic. **HABITAT AND HABITS** Found most often in wet forests and rainforest fringes. Nests underground at base of grass tussock or tree. Workers forage for other insects and honeydew, and can be aggressive, particularly when defending their nest and territory. Capable of stinging humans, jumping forwards and gripping the skin with its jaws while stinging with its abdomen, often doing so several times in quick succession. Venom causes intense localized burning pain and swelling, capable of producing severe allergic reactions, and has been responsible for human fatalities.

Kristi Ellingsen

Black House Ant ■ *Ochetellus glaber* TL 2–4mm

DESCRIPTION Typically brown to black with yellowish antennae, mandibles and legs. Subspecies *clarithorax* (from Brisbane, Qld, to around Sydney, NSW) more reddish-brown with reddish-yellow on thorax and parts of legs. **DISTRIBUTION** Native to Australia, occuring from eastern Qld and NSW (including Lord Howe and Norfolk Islands), to south-western WA. Also introduced to NZ, Solomon Islands, China, India, Japan, the Philippines and USA. **HABITAT AND HABITS** Found in natural settings, open grassland and woodland. Nests in trees, under logs and rocks, or in rotting timber, and forages for honeydew, nectar and small insects. Also enters buildings, attracted by sugary substances, and nests in crevices, cavity walls and electrical appliances; regarded as a pest in these situations. Does not sting or inflict a painful bite on humans.

Rachel Whitlock

Green Ant ■ *Oecophylla smaragdina* TL 8–12mm

(Green Tree Ant)

DESCRIPTION Mostly orange with bright green abdomen. Single queen larger than all workers (to 25mm) and greenish-brown; major workers larger than minor workers, and with larger heads and jaws. **DISTRIBUTION** Throughout tropical northern Australia from western Kimberley, through northern NT, to central eastern Qld. Also Southeast Asia. **HABITAT AND HABITS** Occurs in open forests and woodland. Forms nests in trees by rolling leaves together and connecting them with silk produced by larvae. These are carried by adults and squeezed gently in their jaws to secrete and glue silk to leaves. Same colony can have several nests in adjacent trees. Feeds on honeydew produced by scale insects, which are farmed by minor workers in trees and protected from predators by major workers. Aggressive when disturbed, biting with powerful jaws and spraying formic acid from abdomen.

Peter Rowland/kapeimages.com.au

Peter Rowland/kapeimages.com.au

LEFT Foraging adults; RIGHT Nest.

Coastal Brown Ant ■ *Pheidole megacephala* TL 2–3.5mm

(African Big-headed Ant)

DESCRIPTION Yellowish-brown to dark brown or blackish. Entire body sparsely covered with long hairs, and antennae have clubbed tips. Minor workers small and most often sighted; major workers (about 1 per cent of foraging workers) about twice the size and have enlarged head and powerful jaws. **DISTRIBUTION** Introduced into Australia, with origin most likely in Africa. Widespread in eastern Australia, mainly along coast from around Sydney, NSW, to southern Cape York Peninsula, Qld. **HABITAT AND HABITS** Forms nests in soil, typically in disturbed areas, displacing soil as it excavates below the surface. Omnivorous scavenger of living and dead animal and plant matter underground, in and around buildings, and on trees, readily catching and killing invertebrates, as well as harvesting seeds and honeydew. Considered a major pest in Australia and a significant danger to native ant species, as well as other animal and plant species; classed as one of the world's most invasive species. A single colony has multiple queens.

Sarefa (cc)

Golden-tailed Spiny Ant ■ *Polyrhachis (Hagiomyrma) ammon* TL 6–9mm

DESCRIPTION Black, with most parts of body covered in numerous short to medium golden hairs, generally longer and most noticeable on abdomen. Spines present on mesosoma and petiole. **DISTRIBUTION** Primarily eastern and south-eastern Australia, from north-eastern Qld, through eastern NSW (including Lord Howe Island), to western Vic. **HABITAT AND HABITS** Inhabits open forests and woodland. Nests in sandy soil at bases of trees and shrubs, or under rocks or logs. Omnivorous, feeding on small invertebrates, fruits and nectar, and collecting honeydew from insects such as leafhoppers and scale insects. Not aggressive towards humans, typically retreating or dropping to the ground if disturbed.

Andrew Allen

Judy Gallagher (cc)

Red Fire Ant ■ *Solenopsis invicta* TL 6mm

DESCRIPTION Coppery-brown, darker on abdomen; antennae have distinctive, 2-segmented, clubbed terminal ends. Members of a single colony vary greatly in size. **DISTRIBUTION** Introduced from South America, and recorded in central and south-east Qld, and Sydney region, NSW. **HABITAT AND HABITS** Found in a variety of open disturbed habitats in hot environments. Can form large colonies of up to 400,000 individuals. Feeds on invertebrates, vertebrates (including carrion) and plants. Aggressive and can give a very painful sting, with a burning sensation that can last for up to an hour; itchy pustules may form around bite site that can last for up to 10 days. Can cause severe allergic reactions and anaphylaxis in some individuals.

Ichneumonidae (Ichneumon Wasps)

Striped ichneumon wasps ■ *Gotra* spp. TL 15mm

Peter Rowland/kapeimages.com.au

DESCRIPTION Predominantly black with broad white bands on antennae. Head, body and legs normally have white or yellow bands and spots, and legs have orange-brown bases. Female's ovipositor less than 50 per cent of body length, much shorter than that of the similar **Banded Caterpillar Parasite Wasp** *Ichneumon promissorius*. **DISTRIBUTION** About 8 species occur through eastern Australia, from central eastern Qld, to eastern NSW, Vic, Tas and south-eastern SA. **HABITAT AND HABITS** Found in forests, woodland and gardens, where host species occur. Female lays eggs in cocoons of large moths, using ovipositor, and hatched larvae feed on pupating caterpillar. Adults diurnal and feed on flowers.

Orchid Dupe Wasp ■ *Lissopimpla excelsa* TL 20–25mm
(Dusky Winged Ichneumonid)

DESCRIPTION Adult stout-bodied, orange wasp with blackish wings and pattern of white spots on upper surface of abdomen. Antennae equal to or longer than body, and ovipositor (female) as long as or shorter than body length. **DISTRIBUTION** All states of Australia, although most records from southern states. Also NZ, where probably introduced. **HABITAT AND HABITS** Found in a variety of habitats, including grassland and gardens. Pollinates orchids of the *Cryptostylus* genus, and male is known to mistake parts of flower for female due to mimicry (smell and form) by plant. Female lays eggs in paralysed bodies of armyworm and *Helicoverpa* moth larvae; once hatched wasp larvae feed on and kill their helpless hosts.

John Tann (cc)

Red Ichneumon Wasp ■ *Netelia producta* TL 18–20mm
(Red Devil, Orange Caterpillar Parasite Wasp)

DESCRIPTION Slender, orange-brown wasp with long, flattened abdomen. Wings transparent with black veins. **DISTRIBUTION** Most commonly encountered through eastern and south-eastern Australia. Also NZ. **HABITAT AND HABITS** Found in various vegetated habitats. Eggs laid on armyworms and *Helicoverpa* caterpillars, with caterpillar temporarily paralysed while female attaches a single egg to skin of host, just behind head. Once hatched larva stays attached to caterpillar until its host burrows into soil to pupate. During the caterpillar's pupation stage the wasp larva burrows inside its host and kills it, forming a cocoon within which to pupate itself.

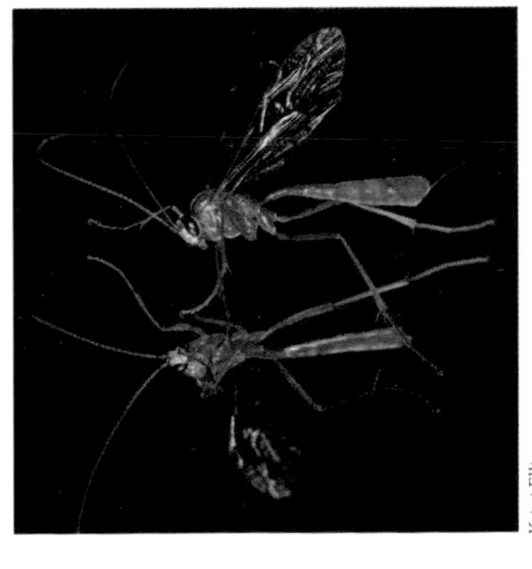

Kristi Ellingsen

MEGACHILIDAE (LEAFCUTTER BEES)

Leafcutter bees ■ *Megachile* spp. TL 6–15mm

DESCRIPTION Adults typically black with longer white hairs. Abdomen wide, tapering to a point, with hairy white stripes on upper surface and orange underside, or entirely orange. Powerful mandibles for cutting soft leaves. **DISTRIBUTION** Throughout Australia. **HABITAT AND HABITS** Occurs in woodland, shrubland and vegetated gardens, where female cuts distinctive circular and oval pieces from leaves to make cells of nest. Nest often in crevices, including hollow stems and crevices in tree bark, but can also be in door and window frames, or in garden hoses and similar garden equipment. Pieces of leaves are used to line crevice and egg is laid in each cell; store of nectar and pollen provided for larva once hatched. Multiple cells are constructed in front of each other until crevice is filled. Some species burrow in soil.

Peter Rowland/kapeimages.com.au

POMPILIDAE (SPIDER WASPS)

Tarantula Hawk Wasp ■ *Cryptocheilus bicolor* TL 35mm

DESCRIPTION Blackish body with broad, dark yellow-orange tip on abdomen, orange-brown wings (without black tips), yellow-orange head and long yellow antennae. **DISTRIBUTION** Throughout Australia in suitable habitats. **HABITAT AND HABITS** Found in forests, woodland, wetlands, heaths and adjacent urban areas, where adults feed on nectar and fruits. Female hunts for large huntsman and wolf spiders as food for larvae after they hatch, paralysing a spider and dragging it to a burrow before laying a single egg on it. Neither male nor female are aggressive towards humans, and rarely sting. If they do, however, the sting has been described as excruciatingly painful.

Jenny Thynne

Peter Rowland/kapeimages.com.au

LEFT Adult at rest; RIGHT Adult female with paralysed spider.

Sphecidae (Mud Daubers)

Mud Dauber ■ *Sceliphron laetum* TL 17–25mm

DESCRIPTION Body yellowish to orange and black. First abdominal segment long and slender, forming thin, waist-like petiole. **DISTRIBUTION** Suitable habitats throughout Australia. Wider spread in Pacific, where it is both naturally occurring and introduced. **HABITAT AND HABITS** Occurs in a variety of wooded and urban habitats, where adults feed on nectar and honeydew. Female gathers wet clay to construct nests that contain several cells, each of which contains an egg and a paralysed invertebrate such as a spider or caterpillar, on which the larva feed once hatched. After consuming its meal, larva pupates within the mud cell before emerging.

Peter Rowland/kapeimages.com.au

Tiphiidae (Tiphiid Wasps)

Blue Ant ■ *Diamma bicolor* TL 30mm

DESCRIPTION Female has metallic blue or greenish body and head, and orange-brown legs and antennae, and is wingless. Male winged and blackish, with incomplete whitish bands on abdomen. **DISTRIBUTION** Throughout south-eastern Australia in suitable habitats, from south-eastern Qld, through eastern NSW and Vic, to Tas. **HABITAT AND HABITS** Found in forests, woodland and urban parks and gardens, where it feeds mostly on nectar. Flying male picks up female and mates with her in flight, after which she constructs a burrow in which to lay her eggs. Sting of female can cause intense pain, swelling and allergic reactions in humans.

Kristi Ellingsen

Kristi Ellingsen

LEFT Female; RIGHT Male.

Vespidae (Vespid Wasps)

Potter Wasp ■ *Abispa (Abispa) ephippium* TL 30mm

DESCRIPTION Orange-yellow with black markings, including black triangle on thorax and broad black band on abdomen. Wings brownish, tipped blackish. **DISTRIBUTION** Throughout mainland Australia. **HABITAT AND HABITS** Common around gardens and similarly wooded areas; mostly seen foraging around foliage or on patches of bare soil. Solitary. Nest has several separate cells made from wet mud, attached to trunk of tree or wall of building, with an egg laid in each cell. Paralysed caterpillars are stored in cells for larvae to feed on once hatched, and female repeatedly brings fresh caterpillars. During construction phase, nest has a funnel-shaped entrance that is sealed off when a new cell is started.

Andrew Allen

Paper Wasps ■ *Polistes* spp. TL 10–25mm

DESCRIPTION Predominantly black with 2 pairs of brownish wings, and scattered yellow or orange markings or rings on long, cylindrical abdomen. Very narrow waist and small head with large eyes. **DISTRIBUTION** Throughout Australia. **HABITAT AND HABITS** Occurs in a variety of habitats throughout range, including urban areas, forests and woodland, where it forms small colonies that build distinctive papery nests from a mixture of wood shavings and saliva. Adults aggressively defend their nest. Sting injects a venom that causes intense pain and swelling around the immediate area, and the victim can be stung multiple times by the same wasp.

Peter Rowland/kapeimages.com.au

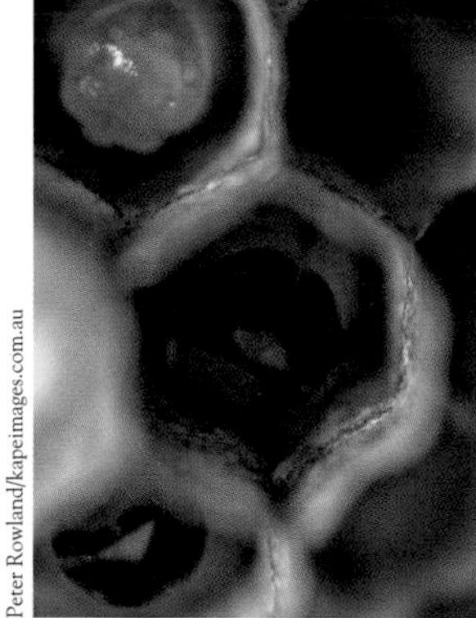

Peter Rowland/kapeimages.com.au

LEFT Adult guarding nest; RIGHT Developing larvae.

European Wasp ■ *Vespula germanica* TL 15mm

(Yellowjacket)

DESCRIPTION Black with broad yellow bands, and almost hairless, elongated abdomen. **DISTRIBUTION** Introduced into southern Australia, where it is widespread from southern Qld in the east to Perth, WA, in the west, and in Tas. Native to Europe, North Africa and Asia. **HABITAT AND HABITS** Found in cool, moist areas, where it builds a papery nest in a crevice or tree hollow, generally close to the ground, but nests have also been found in the roof spaces of dwellings. Food consists of insects, carrion and sweet, sugary substances, including fruits and honeydew, as well as manufactured cakes and sugary drinks. Stings cause pain, inflammation and itchiness, and allergic reactions in some people.

Peter Rowland/kapeimages.com.au

Pergidae (Pergid Sawflies)

Bottlebrush Sawfly ■ *Pterygophorus cinctus* TL 40mm

DESCRIPTION Larvae mostly hairless, with blackish-brown head that becomes paler and reddish with age, and stout reddish, orange, greenish and yellowish body, ending in long, narrow, dark tip. **DISTRIBUTION** Throughout south-eastern Australia in suitable habitats, from south-eastern Qld, through eastern NSW, to southern Vic and Tas. **HABITAT AND HABITS** Found in habitats where bottlebrushes grow. Female lays her eggs inside stems and leaves. After hatching, young larvae are gregarious, feeding together and skeletonizing leaves of the tree they were laid on, but feeding on separate leaves when they are larger. Once fully grown, they drop to the leaf litter and pupate without a cocoon. If handled, larvae may secrete an irritating liquid.

Lou Wolfers

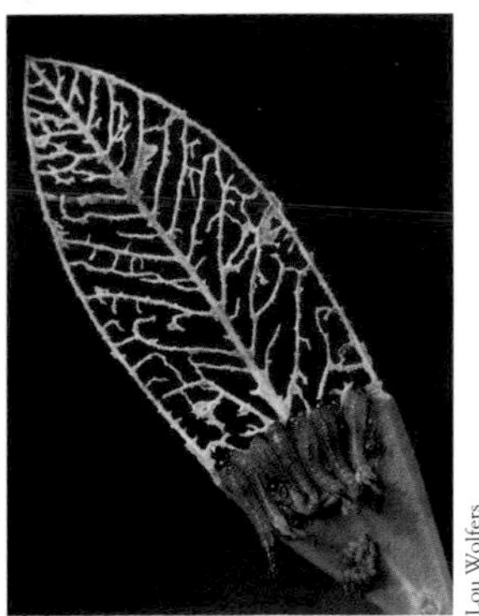

Lou Wolfers

LEFT Adult; RIGHT Larvae.

The following checklist of Australian insects has been developed using data supplied by the Australian Faunal Database (www.biodiversity.org.au/afd/home, supplied 19th November 2021). The total number of currently described genera, species and subspecies for each family represented in Australia has been provided.

KEY TO SYMBOLS
* Representative genus/species accounts included in this book.
ø Taxa within orders undergo ametybolous development (i.e. lacking metamorphsis).
Ŧ Taxa within orders undergo incomplete metamorphosis (i.e. nymphs and adult).
ǂ Taxa within orders undergo complete metamorphosis (i.e. larva, pupa and adult).

AUSTRALIAN INSECTS (CLASS: INSECTA)

Subclass, Order and Family		Genera	Species
Subclass: Apterygotes [Wingless (primitive) Insects]			
ORDER: ARCHAEOGNATHA [Bristletails]ø			
	FAMILY: MEINERTELLIDAE - Rock Bristletails *	5	10
ORDER: ZYGENTOMA (= Thysanura) [Silverfish]ø			
	FAMILY: LEPISMATIDAE *	11	40
	FAMILY: NICOLETIIDAE	14	36
Subclass: Pterygotes [Winged Insects]			
ORDER: ODONATA [Dragonflies and Damselflies]Ŧ			
	FAMILY: COENAGRIONIDAE *	13	31
	FAMILY: ISOSTICTIDAE	7	16
	FAMILY: PLATYCNEMIDIDAE *	1	12
	FAMILY: SYNLESTIDAE	3	7
	FAMILY: HEMIPHLEBIIDAE	1	1
	FAMILY: LESTIDAE *	3	14
	FAMILY: CALOPTERYGIDAE	1	1
	FAMILY: CHLOROCYPHIDAE	1	1
	FAMILY: LESTOIDEIDAE	2	9
	FAMILY: MEGAPODAGRIONIDAE *	5	22
	FAMILY: GOMPHIDAE *	6	35
	FAMILY: LINDENIIDAE *	1	3
	FAMILY: PETALURIDAE	1	5
	FAMILY: BRACHYTRONIDAE	1	1
	FAMILY: AESHNIDAE *	6	13
	FAMILY: AUSTROPETALIIDAE	2	4
	FAMILY: TELEPHLEBIIDAE	8	36
	FAMILY: AUSTROCORDULIIDAE	6	12
	FAMILY: CORDULEPHYIDAE	1	4
	FAMILY: CORDULIIDAE *	4	13
	FAMILY: GOMPHOMACROMIIDAE	1	2
	FAMILY: LIBELLULIDAE *	27	59
	FAMILY: MACROMIIDAE	1	2
	FAMILY: PSEUDOCORDULIIDAE	1	2
	FAMILY: SYNTHEMISTIDAE	8	26
ORDER: PHASMIDA [Stick Insects and Leaf Insects]Ŧ			
	FAMILY: DIAPHEROMERIDAE	11	32
	FAMILY: PHASMATIDAE *	33	65
	FAMILY: PHYLLIIDAE *	3	3

Subclass, Order and Family		Genera	Species
ORDER: TRICHOPTERA [Caddisflies]			
	FAMILY: DIPSEUDOPSIDAE	1	1
	FAMILY: ECNOMIDAE	7	123
	FAMILY: POLYCENTROPODIDAE	5	18
	FAMILY: PSYCHOMYIIDAE	2	3
	FAMILY: PHILOPOTAMIDAE	2	82
	FAMILY: STENOPSYCHIDAE	1	9
	FAMILY: HYDROPSYCHIDAE *	9	55
	FAMILY: ATRIPLECTIDIDAE	1	2
	FAMILY: CALAMOCERATIDAE *	1	10
	FAMILY: LEPTOCERIDAE *	15	176
	FAMILY: ODONTOCERIDAE	2	11
	FAMILY: PHILORHEITHRIDAE	5	15
	FAMILY: ANTIPODOECIIDAE	1	1
	FAMILY: CALOCIDAE	6	32
	FAMILY: CHATHAMIIDAE	1	1
	FAMILY: CONOESUCIDAE	6	23
	FAMILY: HELOCCABUCIDAE	1	1
	FAMILY: HELICOPHIDAE	2	7
	FAMILY: HELICOPSYCHIDAE	1	15
	FAMILY: TASIMIIDAE	2	7
	FAMILY: PLECTROTARSIDAE	3	5
	FAMILY: LIMNEPHILIDAE	1	2
	FAMILY: OECONESIDAE	1	1
	FAMILY: KOKIRIIDAE	3	5
	FAMILY: GLOSSOSOMATIDAE	1	23
	FAMILY: HYDROPTILIDAE	15	157
	FAMILY: HYDROBIOSIDAE	15	65
ORDER: MANTODEA [Praying Mantises (Mantids)]Ŧ			
	FAMILY: AMORPHOSCELIDAE *	8	27
	FAMILY: HYMENOPODIDAE	1	1
	FAMILY: LITURGUSIDAE *	1	9
	FAMILY: MANTIDAE *	27	73
ORDER: EPHEMEROPTERA [Mayflies]Ŧ			
	FAMILY: AMELETOPSIDAE	1	3
	FAMILY: LEPTOPHLEBIIDAE	21	75
	FAMILY: BAETIDAE	6	19
	FAMILY: CAENIDAE	3	12
	FAMILY: COLOBURISCIDAE	1	3
	FAMILY: ONISCIGASTRIDAE	1	3
	FAMILY: PROSOPISTOMATIDAE	1	2
	FAMILY: NESAMELETIDAE	1	1
	FAMILY: VIETNAMELLIDAE	1	1
ORDER: PLECOPTERA [Stoneflies]Ŧ			
	FAMILY: AUSTROPERLIDAE	5	10
	FAMILY: GRIPOPTERYGIDAE *	14	137
	FAMILY: EUSTHENIIDAE *	3	15
	FAMILY: NOTONEMOURIDAE	6	31
ORDER: BLATTODEA [Cockroaches and Termites]Ŧ			
	FAMILY: CORYDIIDAE	3	5
	FAMILY: NOCTICOLIDAE	2	8
	FAMILY: BLABERIDAE *	15	89
	FAMILY: ECTOBIIDAE *	6	40
	FAMILY: TRYONICIDAE	1	3

Subclass, Order and Family		Genera	Species
	FAMILY: BLATTIDAE *	22	197
	FAMILY: KALOTERMITIDAE	9	35
	FAMILY: RHINOTERMITIDAE *	5	25
	FAMILY: TERMITIDAE *	24	190
	FAMILY: MASTOTERMITIDAE *	1	1
	FAMILY: TERMOPSIDAE	2	5
	FAMILY: BLATTELLIDAE*	19	148
	FAMILY: PSEUDOPHYLLODROMIDAE	8	65
ORDER: EMBIOPTERA [Web-spinners]Ŧ			
	FAMILY: AUSTRALEMBIIDAE	1	18
	FAMILY: NOTOLIGOTOMIDAE	1	2
	FAMILY: OLIGOTOMIDAE	2	6
ORDER: ORTHOPTERA [Grasshoppers, Crickets and Katydids]Ŧ			
	FAMILY: ACRIDIDAE *	135	249
	FAMILY: PYRGOMORPHIDAE *	12	26
	FAMILY: MORABIDAE	42	96
	FAMILY: TETRIGIDAE	14	38
	FAMILY: CYLINDRACHETIDAE	2	14
	FAMILY: TRIDACTYLIDAE	3	8
	FAMILY: ANOSTOSTOMATIDAE *	6	13
	FAMILY: COOLOOLIDAE	1	4
	FAMILY: GRYLLACRIDIDAE	29	91
	FAMILY: RHAPHIDOPHORIDAE	13	25
	FAMILY: GRYLLIDAE *	48	310
	FAMILY: MOGOPLISTIDAE	11	85
	FAMILY: PHALANGOPSIDAE	14	41
	FAMILY: TRIGONIDIIDAE	21	79
	FAMILY: TETTIGONIIDAE *	104	378
	FAMILY: GRYLLOTALPIDAE *	2	12
	FAMILY: MYRMECOPHILIDAE	1	7
ORDER: DERMAPTERA [Earwigs]Ŧ			
	FAMILY: ANISOLABIDIDAE	10	31
	FAMILY: APACHYIDAE	1	4
	FAMILY: CHELISOCHIDAE	4	9
	FAMILY: FORFICULIDAE *	5	6
	FAMILY: LABIDURIDAE *	3	3
	FAMILY: PYGIDICRANIDAE	5	9
	FAMILY: SPONGIPHORIDAE	10	27
ORDER: ZORAPTERA [Zoropterans]Ŧ			
	FAMILY: ZOROTYPIDAE	1	1
ORDER: PSOCODEA (= Psocoptera + Phthiraptera) [Booklice, Barklice, Biting and Sucking (True) Lice]Ŧ			
	FAMILY: ARCHIPSOCIDAE *	2	3
	FAMILY: ELIPSOCIDAE	18	25
	FAMILY: ECTOPSOCIDAE	1	34
	FAMILY: HEMIPSOCIDAE	1	2
	FAMILY: MESOPSOCIDAE	1	1
	FAMILY: MYOPSOCIDAE	5	12
	FAMILY: PERIPSOCIDAE	2	21
	FAMILY: PHILOTARSIDAE	4	23
	FAMILY: PSEUDOCAECILIIDAE	13	38
	FAMILY: PSILOPSOCIDAE	1	1
	FAMILY: PSOCIDAE *	12	58
	FAMILY: TRICHOPSOCIDAE	1	1
	FAMILY: EPIPSOCIDAE	1	1

Subclass, Order and Family		Genera	Species
	FAMILY: CAECILIUSIDAE	14	42
	FAMILY: STENOPSOCIDAE	1	1
	FAMILY: AMPHIPSOCIDAE	1	3
	FAMILY: LACHESILLIDAE	1	2
	FAMILY: CALOPSOCIDAE	1	3
	FAMILY: AMPHIENTOMIDAE	4	5
	FAMILY: LIPOSCELIDIDAE *	2	6
	FAMILY: PACHYTROCTIDAE	2	5
	FAMILY: SPHAEROPSOCIDAE	1	1
	FAMILY: LEPIDOPSOCIDAE	10	30
	FAMILY: TROGIIDAE	3	10
	FAMILY: PSOQUILLIDAE	3	3
	FAMILY: PSYLLIPSOCIDAE	3	4
	FAMILY: BOOPIIDAE	6	44
	FAMILY: GYROPIDAE	2	2
	FAMILY: LAEMOBOTHRIIDAE	1	4
	FAMILY: MENOPONIDAE *	29	112
	FAMILY: RICINIDAE	1	6
	FAMILY: ECHINOPHTHIRIIDAE	2	5
	FAMILY: HOPLOPLEURIDAE	1	13
	FAMILY: HAEMATOPINIDAE *	1	5
	FAMILY: LINOGNATHIDAE	2	6
	FAMILY: PEDICULIDAE *	1	1
	FAMILY: POLYPLACIDAE	2	3
	FAMILY: PTHIRIDAE *	1	1
	FAMILY: PHILOPTERIDAE	67	277
	FAMILY: TRICHODECTIDAE *	4	7
ORDER: THYSANOPTERA [Thrips]Ŧ			
	FAMILY: AEOLOTHRIPIDAE	10	31
	FAMILY: THRIPIDAE *	85	272
	FAMILY: MELANTHRIPIDAE	2	13
	FAMILY: MEROTHRIPIDAE	2	5
	FAMILY: UZELOTHRIPIDAE	1	1
	FAMILY: PHLAEOTHRIPIDAE *	142	586
ORDER: HEMIPTERA [True Bugs]Ŧ			
	FAMILY: CLASTOPTERIDAE	5	10
	FAMILY: CERCOPIDAE *	14	29
	FAMILY: CICADIDAE *	88	325
	FAMILY: TETTIGARCTIDAE	1	2
	FAMILY: CICADELLIDAE *	235	636
	FAMILY: MEMBRACIDAE *	37	81
	FAMILY: ACHILIDAE	23	36
	FAMILY: CIXIIDAE	53	213
	FAMILY: DICTYOPHARIDAE	8	15
	FAMILY: NOGODINIDAE	11	19
	FAMILY: CALISCELIDAE	1	1
	FAMILY: DELPHACIDAE	51	85
	FAMILY: DERBIDAE	19	52
	FAMILY: EURYBRACHIDAE	16	55
	FAMILY: FLATIDAE *	28	87
	FAMILY: FULGORIDAE	7	20
	FAMILY: ISSIDAE	5	12
	FAMILY: LOPHOPIDAE	2	3
	FAMILY: MEENOPLIDAE	2	14

Subclass, Order and Family		Genera	Species
	FAMILY: RICANIIDAE *	9	29
	FAMILY: TROPIDUCHIDAE	13	19
	FAMILY: PELORIDIIDAE	8	12
	FAMILY: ADELGIDAE	2	4
	FAMILY: APHIDIDAE *	83	160
	FAMILY: PHYLLOXERIDAE	3	3
	FAMILY: PSEUDOCOCCIDAE *	66	193
	FAMILY: ASTEROLECANIIDAE	7	27
	FAMILY: ERIOCOCCIDAE *	25	170
	FAMILY: DIASPIDIDAE	89	240
	FAMILY: COCCIDAE *	26	84
	FAMILY: ACLERDIDAE	1	1
	FAMILY: CALLIPAPPIDAE	2	6
	FAMILY: MARGARODIDAE	2	2
	FAMILY: MONOPHLEBIDAE *	5	19
	FAMILY: STEINGELLIIDAE	2	4
	FAMILY: BEESONIIDAE	1	1
	FAMILY: CEROCOCCIDAE	1	6
	FAMILY: CONCHASPIDIDAE	1	1
	FAMILY: DACTYLOPIIDAE *	1	5
	FAMILY: HALIMOCOCCIDAE	1	2
	FAMILY: KERRIIDAE	3	7
	FAMILY: LECANODIASPIDIDAE	4	19
	FAMILY: ORTHEZIIDAE	4	9
	FAMILY: TRIOZIDAE	10	64
	FAMILY: APHALARIDAE	1	1
	FAMILY: PSYLLIDAE *	32	314
	FAMILY: CALOPHYIDAE	3	6
	FAMILY: CARSIDARIDAE	5	7
	FAMILY: HOMOTOMIDAE	1	2
	FAMILY: LIVIIDAE	5	8
	FAMILY: PHACOPTERONIDAE	1	2
	FAMILY: ALEYRODIDAE *	45	115
	FAMILY: ANTHOCORIDAE	17	32
	FAMILY: CIMICIDAE *	1	1
	FAMILY: NABIDAE	8	25
	FAMILY: PLOKIOPHILIDAE	1	1
	FAMILY: POLYCTENIDAE	2	2
	FAMILY: THAUMASTOCORIDAE	3	21
	FAMILY: MIRIDAE	197	529
	FAMILY: REDUVIIDAE *	103	221
	FAMILY: TINGIDAE *	61	213
	FAMILY: GERRIDAE *	11	37
	FAMILY: VELIIDAE	12	65
	FAMILY: HERMATOBATIDAE	1	2
	FAMILY: HEBRIDAE	3	6
	FAMILY: HYDROMETRIDAE	1	8
	FAMILY: MESOVELIIDAE	2	7
	FAMILY: CERATOCOMBIDAE	1	1
	FAMILY: SCHIZOPTERIDAE	18	100
	FAMILY: DIPSOCORIDAE	1	4
	FAMILY: AENICTOPECHEIDAE	2	2
	FAMILY: ENICOCEPHALIDAE	3	5
	FAMILY: LEPTOPODIDAE	1	2

Subclass, Order and Family		Genera	Species
	FAMILY: OMANIIDAE	1	1
	FAMILY: SALDIDAE	3	10
	FAMILY: CORIXIDAE *	4	17
	FAMILY: MICRONECTIDAE	2	17
	FAMILY: NAUCORIDAE	4	8
	FAMILY: BELOSTOMATIDAE *	2	4
	FAMILY: NEPIDAE	5	9
	FAMILY: NOTONECTIDAE *	6	42
	FAMILY: PLEIDAE	1	3
	FAMILY: GELASTOCORIDAE	1	25
	FAMILY: OCHTERIDAE	2	10
	FAMILY: ARADIDAE	38	128
	FAMILY: TERMITAPHIDIDAE	1	1
	FAMILY: ARTHENEIDAE	1	2
	FAMILY: GEOCORIDAE	8	25
	FAMILY: BERYTIDAE	6	7
	FAMILY: BLISSIDAE	9	15
	FAMILY: COLOBATHRISTIDAE	1	1
	FAMILY: CRYPTORHAMPHIDAE	2	4
	FAMILY: CYMIDAE	4	10
	FAMILY: HETEROGASTRIDAE	4	6
	FAMILY: LYGAEIDAE *	23	78
	FAMILY: MESCHIIDAE	2	3
	FAMILY: NINIDAE	2	2
	FAMILY: OXYCARENIDAE	1	4
	FAMILY: PACHYGRONTHIDAE	6	10
	FAMILY: PIESMATIDAE	1	11
	FAMILY: RHYPAROCHROMIDAE	75	171
	FAMILY: ALYDIDAE *	7	16
	FAMILY: COREIDAE *	46	91
	FAMILY: HYOCEPHALIDAE	2	3
	FAMILY: RHOPALIDAE	2	7
	FAMILY: STENOCEPHALIDAE	1	1
	FAMILY: HENICOCORIDAE	1	1
	FAMILY: IDIOSTOLIDAE	2	3
	FAMILY: ACANTHOSOMATIDAE	17	46
	FAMILY: APHYLIDAE	2	3
	FAMILY: CYDNIDAE	21	75
	FAMILY: DINIDORIDAE	4	6
	FAMILY: LESTONIIDAE	1	2
	FAMILY: PENTATOMIDAE *	136	337
	FAMILY: PLATASPIDAE	2	20
	FAMILY: SCUTELLERIDAE *	13	27
	FAMILY: TESSARATOMIDAE	12	18
	FAMILY: LARGIDAE	2	4
	FAMILY: PYRRHOCORIDAE	4	12
ORDER: NEUROPTERA [Lacewings, Antlions and Mantidflies]‡			
	FAMILY: ASCALAPHIDAE *	15	36
	FAMILY: BEROTHIDAE	8	26
	FAMILY: CHRYSOPIDAE *	16	58
	FAMILY: CONIOPTERYGIDAE	5	41
	FAMILY: HEMEROBIIDAE	10	34
	FAMILY: ITHONIDAE	3	14
	FAMILY: MANTISPIDAE *	10	48

Subclass, Order and Family		Genera	Species
	FAMILY: MYRMELEONTIDAE *	43	211
	FAMILY: NEMOPTERIDAE	3	9
	FAMILY: NEURORTHIDAE	1	1
	FAMILY: NYMPHIDAE *	7	21
	FAMILY: OSMYLIDAE	11	30
	FAMILY: PSYCHOPSIDAE	1	13
	FAMILY: SISYRIDAE	2	9
ORDER: MEGALOPTERA [Alderflies, Dobsonflies and Fishflies]‡			
	FAMILY: CORYDALIDAE *	3	22
	FAMILY: SIALIDAE *	2	4
ORDER: COLEOPTERA [Beetles]‡			
	FAMILY: CUPEDIDAE	2	6
	FAMILY: OMMATIDAE	2	5
	FAMILY: SPHAERIUSIDAE	1	2
	FAMILY: CARABIDAE *	323	2026
	FAMILY: DYTISCIDAE *	40	299
	FAMILY: GYRINIDAE *	4	18
	FAMILY: HALIPLIDAE	1	18
	FAMILY: HYGROBIIDAE	1	4
	FAMILY: NOTERIDAE	4	6
	FAMILY: RHYSODIDAE	5	13
	FAMILY: HISTERIDAE	50	246
	FAMILY: HYDROPHILIDAE	44	196
	FAMILY: HYDRAENIDAE	10	209
	FAMILY: PTILIIDAE	20	56
	FAMILY: LEIODIDAE	22	106
	FAMILY: SILPHIDAE	2	3
	FAMILY: STAPHYLINIDAE *	465	1552
	FAMILY: DERMESTIDAE *	18	132
	FAMILY: JACOBSONIIDAE	2	2
	FAMILY: NOSODENDRIDAE	1	2
	FAMILY: BOSTRICHIDAE	24	58
	FAMILY: PTINIDAE	38	190
	FAMILY: ANTHRIBIDAE	49	78
	FAMILY: ATTELABIDAE	3	83
	FAMILY: BELIDAE	18	115
	FAMILY: BRENTIDAE	58	162
	FAMILY: CURCULIONIDAE *	706	2215
	FAMILY: CARIDAE	2	4
	FAMILY: NEMONYCHIDAE	8	16
	FAMILY: LUCANIDAE *	19	89
	FAMILY: TROGIDAE	2	56
	FAMILY: GEOTRUPIDAE	1	1
	FAMILY: HYBOSORIDAE	5	45
	FAMILY: PASSALIDAE *	9	35
	FAMILY: SCARABAEIDAE *	283	1729
	FAMILY: BOLBOCERATIDAE	10	161
	FAMILY: CERAMBYCIDAE *	254	983
	FAMILY: CHRYSOMELIDAE *	235	1807
	FAMILY: MEGALOPODIDAE	2	4
	FAMILY: CLAMBIDAE	3	21
	FAMILY: SCIRTIDAE	48	282
	FAMILY: BRACHYPSECTRIDAE	1	1
	FAMILY: DASCILLIDAE	1	3

Subclass, Order and Family		Genera	Species
	FAMILY: RHIPICERIDAE	1	6
	FAMILY: BUPRESTIDAE *	80	1066
	FAMILY: BYRRHIDAE	7	33
	FAMILY: PTILODACTYLIDAE	3	6
	FAMILY: CALLIRHIPIDAE	2	4
	FAMILY: CHELONARIIDAE	1	1
	FAMILY: ELMIDAE	9	102
	FAMILY: HETEROCERIDAE	2	8
	FAMILY: LIMNICHIDAE	4	9
	FAMILY: PSEPHENIDAE	1	13
	FAMILY: CANTHARIDAE *	3	115
	FAMILY: ELATERIDAE *	69	563
	FAMILY: EUCNEMIDAE	36	86
	FAMILY: LAMPYRIDAE *	8	25
	FAMILY: LYCIDAE *	13	199
	FAMILY: RHINORHIPIDAE	1	1
	FAMILY: THROSCIDAE	3	10
	FAMILY: DERODONTIDAE	1	1
	FAMILY: EUCINETIDAE	2	14
	FAMILY: LYMEXYLIDAE	6	7
	FAMILY: ACANTHOCNEMIDAE	1	1
	FAMILY: CLERIDAE	49	359
	FAMILY: MELYRIDAE *	13	324
	FAMILY: PHYCOSECIDAE	1	3
	FAMILY: TROGOSSITIDAE	16	37
	FAMILY: BOGANIIDAE	3	6
	FAMILY: CERYLONIDAE	7	22
	FAMILY: CUCUJIDAE	2	7
	FAMILY: NITIDULIDAE	59	172
	FAMILY: ANAMORPHIDAE	3	21
	FAMILY: BIPHYLLIDAE	3	20
	FAMILY: BOTHRIDERIDAE	7	32
	FAMILY: CAVOGNATHIDAE	1	2
	FAMILY: COCCINELLIDAE *	61	321
	FAMILY: CYBOCEPHALIDAE	2	4
	FAMILY: CORYLOPHIDAE	10	58
	FAMILY: CRYPTOPHAGIDAE	7	16
	FAMILY: DISCOLOMATIDAE	1	8
	FAMILY: ENDOMYCHIDAE	9	27
	FAMILY: EROTYLIDAE	21	69
	FAMILY: EUXESTIDAE	4	15
	FAMILY: HOBARTIIDAE	2	5
	FAMILY: KATERETIDAE	2	11
	FAMILY: LAEMOPHLOEIDAE	12	58
	FAMILY: LAMINGTONIIDAE	1	3
	FAMILY: LATRIDIIDAE	13	31
	FAMILY: MONOTOMIDAE	5	10
	FAMILY: MURMIDIIDAE	1	1
	FAMILY: MYCTAEIDAE	1	1
	FAMILY: MYRABOLIIDAE	1	13
	FAMILY: PASSANDRIDAE	3	9
	FAMILY: PHALACRIDAE	6	69
	FAMILY: PHLOEOSTICHIDAE	3	5
	FAMILY: PRIASILPHIDAE	1	3

Subclass, Order and Family		Genera	Species
	FAMILY: PROTOCUCUJIDAE	1	3
	FAMILY: SILVANIDAE	18	57
	FAMILY: SPHINDIDAE	2	6
	FAMILY: TASMOSALPINGIDAE	1	2
	FAMILY: TEREDIDAE	4	22
	FAMILY: ADERIDAE	6	78
	FAMILY: CIIDAE	22	94
	FAMILY: MYCETOPHAGIDAE	5	6
	FAMILY: TENEBRIONIDAE *	211	1140
	FAMILY: ULODIDAE	7	14
	FAMILY: ANTHICIDAE	32	234
	FAMILY: ARCHEOCRYPTICIDAE	9	22
	FAMILY: BORIDAE	1	1
	FAMILY: MELANDRYIDAE	8	26
	FAMILY: MELOIDAE	7	50
	FAMILY: MORDELLIDAE	11	132
	FAMILY: MYCTERIDAE	3	15
	FAMILY: OEDEMERIDAE	10	84
	FAMILY: PROMECHEILIDAE	2	3
	FAMILY: PROSTOMIDAE	2	7
	FAMILY: PYROCHROIDAE	4	18
	FAMILY: PYTHIDAE	1	1
	FAMILY: RIPIPHORIDAE	11	67
	FAMILY: SALPINGIDAE	10	25
	FAMILY: SCRAPTIIDAE	2	17
	FAMILY: ZOPHERIDAE	30	100
ORDER: STREPSIPTERA [Twisted Wings and Stylopids]ǂ			
	FAMILY: CORIOXENIDAE	2	8
	FAMILY: ELENCHIDAE	2	4
	FAMILY: HALICTOPHAGIDAE	5	14
	FAMILY: MENGENILLIDAE	1	2
	FAMILY: MYRMECOLACIDAE	3	10
	FAMILY: STYLOPIDAE	2	4
ORDER: DIPTERA [Flies, Mosquitoes and Gnats]ǂ			
	FAMILY: ANISOPODIDAE	2	4
	FAMILY: CECIDOMYIIDAE	50	144
	FAMILY: MYCETOPHILIDAE	25	46
	FAMILY: SCIARIDAE	20	125
	FAMILY: BIBIONIDAE	4	31
	FAMILY: DIADOCIDIIDAE	1	2
	FAMILY: DITOMYIIDAE	1	9
	FAMILY: KEROPLATIDAE	22	43
	FAMILY: VALESEGUYIDAE	1	1
	FAMILY: BLEPHARICERIDAE	5	28
	FAMILY: PSYCHODIDAE	27	146
	FAMILY: PERISSOMMATIDAE	1	4
	FAMILY: SCATOPSIDAE	13	57
	FAMILY: TANYDERIDAE	3	6
	FAMILY: LIMONIIDAE	49	824
	FAMILY: TIPULIDAE *	12	182
	FAMILY: CYLINDROTOMIDAE	1	3
	FAMILY: PEDICIIDAE	1	2
	FAMILY: TRICHOCERIDAE	3	10
	FAMILY: CHAOBORIDAE	3	7

Subclass, Order and Family		Genera	Species
	FAMILY: CORETHRELLIDAE	1	3
	FAMILY: CULICIDAE	18	215
	FAMILY: DIXIDAE	2	6
	FAMILY: CERATOPOGONIDAE *	32	306
	FAMILY: CHIRONOMIDAE *	91	240
	FAMILY: SIMULIIDAE	8	46
	FAMILY: THAUMALEIDAE	2	29
	FAMILY: XYLOPHAGIDAE	1	4
	FAMILY: STRATIOMYIDAE *	36	131
	FAMILY: XYLOMYIDAE	1	1
	FAMILY: ATHERICIDAE	2	12
	FAMILY: TABANIDAE *	27	251
	FAMILY: PELECORHYNCHIDAE	1	31
	FAMILY: AUSTROLEPTIDAE	1	3
	FAMILY: RHAGIONIDAE	3	41
	FAMILY: ACROCERIDAE	5	41
	FAMILY: NEMESTRINIDAE	6	51
	FAMILY: APIOCERIDAE	1	73
	FAMILY: ASILIDAE *	68	345
	FAMILY: BOMBYLIIDAE	59	379
	FAMILY: THEREVIDAE	28	314
	FAMILY: APSILOCEPHALIDAE	1	2
	FAMILY: MYDIDAE	4	42
	FAMILY: SCENOPINIDAE	6	86
	FAMILY: DOLICHOPODIDAE	52	418
	FAMILY: EMPIDIDAE	17	80
	FAMILY: HYBOTIDAE	9	32
	FAMILY: RAGADIDAE	2	10
	FAMILY: IRONOMYIIDAE	1	3
	FAMILY: PHORIDAE	23	137
	FAMILY: PIPUNCULIDAE	11	79
	FAMILY: PLATYPEZIDAE	3	19
	FAMILY: SYRPHIDAE *	46	183
	FAMILY: CONOPIDAE	19	93
	FAMILY: AUSTRALIMYZIDAE	1	4
	FAMILY: AGROMYZIDAE	16	146
	FAMILY: CLUSIIDAE	6	39
	FAMILY: ODINIIDAE	2	1
	FAMILY: FERGUSONINIDAE	1	35
	FAMILY: CALLIPHORIDAE *	19	118
	FAMILY: RHINOPHORIDAE	3	18
	FAMILY: SARCOPHAGIDAE *	11	91
	FAMILY: RHINIIDAE	4	15
	FAMILY: TACHINIDAE *	137	426
	FAMILY: ULURUMYIIDAE	1	1
	FAMILY: OESTRIDAE	4	6
	FAMILY: CHAMAEMYIIDAE	7	34
	FAMILY: HELOSCIOMYZIDAE	5	16
	FAMILY: LAUXANIIDAE *	29	270
	FAMILY: SCIOMYZIDAE	4	10
	FAMILY: COELOPIDAE	10	13
	FAMILY: SEPSIDAE	7	14
	FAMILY: LONCHOPTERIDAE	1	1
	FAMILY: CHYROMYIDAE	1	1

Subclass, Order and Family		Genera	Species
	FAMILY: SPHAEROCERIDAE	29	69
	FAMILY: HETEROMYZIDAE	17	83
	FAMILY: CYPSELOSOMATIDAE	2	4
	FAMILY: MICROPEZIDAE	6	30
	FAMILY: NERIIDAE *	2	2
	FAMILY: PSILIDAE	2	1
	FAMILY: ASTEIIDAE	2	2
	FAMILY: CANACIDAE	15	42
	FAMILY: CHLOROPIDAE	46	290
	FAMILY: MILICHIIDAE	5	21
	FAMILY: NEMINIDAE	1	6
	FAMILY: NEUROCHAETIDAE	2	3
	FAMILY: PERISCELIDIDAE	3	8
	FAMILY: TERATOMYZIDAE	5	7
	FAMILY: LONCHAEIDAE	4	24
	FAMILY: PIOPHILIDAE	2	7
	FAMILY: PYRGOTIDAE	15	61
	FAMILY: TEPHRITIDAE *	74	274
	FAMILY: ULIDIIDAE	3	4
	FAMILY: PLATYSTOMATIDAE *	24	199
	FAMILY: CRYPTOCHETIDAE	2	4
	FAMILY: EPHYDRIDAE	37	93
	FAMILY: CURTONOTIDAE	1	1
	FAMILY: DROSOPHILIDAE	36	254
	FAMILY: ANTHOMYIIDAE	5	11
	FAMILY: MUSCIDAE *	37	195
	FAMILY: FANNIIDAE	3	12
	FAMILY: HIPPOBOSCIDAE	12	26
	FAMILY: NYCTERIBIIDAE	7	29
	FAMILY: STREBLIDAE	3	5
	FAMILY: BRAULIDAE	1	1
ORDER: MECOPTERA [Scorpionflies and Hanging Flies]‡			
	FAMILY: APTEROPANORPIDAE	1	4
	FAMILY: BITTACIDAE *	6	16
	FAMILY: CHORISTIDAE *	3	8
	FAMILY: MEROPEIDAE	1	1
	FAMILY: NANNOCHORISTIDAE	1	4
ORDER: SIPHONAPTERA [Fleas]‡			
	FAMILY: CERATOPHYLLIDAE	3	4
	FAMILY: HYSTRICHOPSYLLIDAE	2	2
	FAMILY: ISCHNOPSYLLIDAE	4	7
	FAMILY: LEPTOPSYLLIDAE	1	1
	FAMILY: MACROPSYLLIDAE	2	2
	FAMILY: PULICIDAE *	5	20
	FAMILY: PYGIOPSYLLIDAE	14	36
	FAMILY: RHOPALOPSYLLIDAE	1	4
	FAMILY: STEPHANOCIRCIDAE	2	8
ORDER: LEPIDOPTERA [Butterlies and Moths]‡			
	FAMILY: LOPHOCORONIDAE	1	6
	FAMILY: AENIGMATINEIDAE	1	1
	FAMILY: PALEOSETIDAE	1	1
	FAMILY: ANOMOSETIDAE	1	1
	FAMILY: HEPIALIDAE	9	140
	FAMILY: NEPTICULIDAE	4	28

Subclass, Order and Family		Genera	Species
	FAMILY: OPOSTEGIDAE	2	18
	FAMILY: HELIOZELIDAE	5	37
	FAMILY: ADELIDAE	2	11
	FAMILY: INCURVARIIDAE	2	13
	FAMILY: PALAEPHATIDAE	2	11
	FAMILY: MICROPTERIGIDAE	3	9
	FAMILY: AGATHIPHAGIDAE	1	1
	FAMILY: ALUCITIDAE	1	6
	FAMILY: TINEODIDAE	9	12
	FAMILY: LYCAENIDAE *	49	157
	FAMILY: NYMPHALIDAE *	40	98
	FAMILY: PAPILIONIDAE *	6	21
	FAMILY: HESPERIIDAE *	36	127
	FAMILY: PIERIDAE *	9	40
	FAMILY: RIODINIDAE	1	1
	FAMILY: GRACILLARIIDAE	26	198
	FAMILY: BUCCULATRICIDAE *	2	27
	FAMILY: DOUGLASIIDAE	1	1
	FAMILY: ERIOCOTTIDAE	1	1
	FAMILY: GALACTICIDAE	2	6
	FAMILY: PSYCHIDAE *	27	186
	FAMILY: ROESLERSTAMMIIDAE	9	23
	FAMILY: TINEIDAE	40	181
	FAMILY: MEESSIIDAE	4	7
	FAMILY: DRYADAULIDAE	1	9
	FAMILY: ARGYRESTHIIDAE	1	1
	FAMILY: GLYPHIPTERIGIDAE	4	58
	FAMILY: HELIODINIDAE	2	4
	FAMILY: LYONETIIDAE	11	40
	FAMILY: PLUTELLIDAE	10	28
	FAMILY: YPONOMEUTIDAE	21	58
	FAMILY: CYCLOTORNIDAE	1	5
	FAMILY: LACTURIDAE	2	24
	FAMILY: EPIPYROPIDAE	3	5
	FAMILY: LIMACODIDAE *	26	70
	FAMILY: ZYGAENIDAE	12	46
	FAMILY: CARPOSINIDAE	7	39
	FAMILY: COPROMORPHIDAE	5	7
	FAMILY: COSSIDAE	19	98
	FAMILY: DUDGEONEIDAE	1	3
	FAMILY: EPERMENIIDAE	6	21
	FAMILY: IMMIDAE	2	13
	FAMILY: MACROPIRATIDAE	1	1
	FAMILY: PTEROPHORIDAE	26	49
	FAMILY: CRAMBIDAE	251	756
	FAMILY: PYRALIDAE	165	466
	FAMILY: BRACHODIDAE	3	23
	FAMILY: CHOREUTIDAE	5	15
	FAMILY: SESIIDAE	10	18
	FAMILY: LASIOCAMPIDAE	12	72
	FAMILY: ANTHELIDAE *	8	73
	FAMILY: EUPTEROTIDAE *	3	6
	FAMILY: BOMBYCIDAE	3	3
	FAMILY: CARTHAEIDAE	1	1

Subclass, Order and Family		Genera	Species
	FAMILY: SATURNIIDAE	6	19
	FAMILY: CASTNIIDAE	1	31
	FAMILY: DREPANIDAE	6	11
	FAMILY: BATRACHEDRIDAE	3	32
	FAMILY: LECITHOCERIDAE	9	49
	FAMILY: BLASTOBASIDAE	1	13
	FAMILY: BLASTODACNIDAE	5	21
	FAMILY: COLEOPHORIDAE	1	13
	FAMILY: COSMOPTERIGIDAE	33	396
	FAMILY: DEPRESSARIIDAE	26	74
	FAMILY: ELACHISTIDAE	5	145
	FAMILY: ETHMIIDAE	1	14
	FAMILY: GELECHIIDAE	47	433
	FAMILY: HYPERTROPHIDAE	11	50
	FAMILY: MOMPHIDAE	1	1
	FAMILY: OECOPHORIDAE	365	2088
	FAMILY: SCYTHRIDIDAE	3	24
	FAMILY: SYMMOCIDAE	1	1
	FAMILY: GEOMETRIDAE *	324	1329
	FAMILY: HYBLAEIDAE	1	4
	FAMILY: OENOSANDRIDAE	4	8
	FAMILY: NOTODONTIDAE *	40	80
	FAMILY: EREBIDAE *	328	903
	FAMILY: EUTELIIDAE	14	36
	FAMILY: NOLIDAE *	51	169
	FAMILY: NOCTUIDAE *	147	491
	FAMILY: SPHINGIDAE	23	74
	FAMILY: SIMAETHISTIDAE	1	2
	FAMILY: THYRIDIDAE	18	55
	FAMILY: TORTRICIDAE	179	645
	FAMILY: URANIIDAE	19	33
ORDER: HYMENOPTERA [Bees, Wasps, Ants and Sawflies]‡			
	FAMILY: APIDAE *	13	164
	FAMILY: MEGACHILIDAE *	6	168
	FAMILY: COLLETIDAE	32	828
	FAMILY: HALICTIDAE	11	377
	FAMILY: STENOTRITIDAE	2	21
	FAMILY: AMPULICIDAE	4	14
	FAMILY: CRABRONIDAE *	46	756
	FAMILY: SPHECIDAE *	9	59
	FAMILY: AGAONIDAE	8	41
	FAMILY: MYMARIDAE	37	260
	FAMILY: PTEROMALIDAE	207	494
	FAMILY: APHELINIDAE	22	244
	FAMILY: AZOTIDAE	1	44
	FAMILY: CHALCIDIDAE	22	171
	FAMILY: ENCYRTIDAE	143	450
	FAMILY: EUCHARITIDAE	19	90
	FAMILY: EULOPHIDAE	117	698
	FAMILY: EUPELMIDAE	13	169
	FAMILY: EURYTOMIDAE	18	132
	FAMILY: LEUCOSPIDAE	1	11
	FAMILY: MEGASTIGMIDAE	10	82
	FAMILY: ORMYRIDAE	1	12

Subclass, Order and Family		Genera	Species
	FAMILY: PERILAMPIDAE	8	34
	FAMILY: SIGNIPHORIDAE	2	16
	FAMILY: TANAOSTIGMATIDAE	2	11
	FAMILY: TETRACAMPIDAE	4	7
	FAMILY: TORYMIDAE	16	54
	FAMILY: TRICHOGRAMMATIDAE	34	140
	FAMILY: BETHYLIDAE	19	61
	FAMILY: CHRYSIDIDAE *	9	66
	FAMILY: DRYINIDAE	18	114
	FAMILY: EMBOLEMIDAE	2	2
	FAMILY: SCLEROGIBBIDAE	1	2
	FAMILY: SCOLEBYTHIDAE	2	2
	FAMILY: AUSTROCYNIPIDAE	1	1
	FAMILY: CYNIPIDAE	1	1
	FAMILY: FIGITIDAE	21	50
	FAMILY: IBALIIDAE	1	2
	FAMILY: LIOPTERIDAE	1	1
	FAMILY: AULACIDAE	2	79
	FAMILY: GASTERUPTIIDAE	3	190
	FAMILY: EVANIIDAE	4	36
	FAMILY: BRACONIDAE	233	668
	FAMILY: TRACHYPETIDAE	3	9
	FAMILY: ICHNEUMONIDAE *	166	445
	FAMILY: PLATYGASTRIDAE	21	45
	FAMILY: SCELIONIDAE	60	610
	FAMILY: AUSTRONIIDAE	1	3
	FAMILY: DIAPRIIDAE	36	146
	FAMILY: MONOMACHIDAE	1	3
	FAMILY: PERADENIIDAE	1	2
	FAMILY: PROCTOTRUPIDAE	7	23
	FAMILY: FORMICIDAE *	109	1227
	FAMILY: MUTILLIDAE	10	195
	FAMILY: TIPHIIDAE *	55	587
	FAMILY: POMPILIDAE *	46	257
	FAMILY: RHOPALOSOMATIDAE	1	17
	FAMILY: SCOLIIDAE	11	20
	FAMILY: VESPIDAE *	39	342
	FAMILY: PERGIDAE *	24	132
	FAMILY: ARGIDAE	4	11
	FAMILY: ZENARGIDAE	1	1
	FAMILY: TENTHREDINIDAE	9	9
	FAMILY: SIRICIDAE	2	2
	FAMILY: XIPHYDRIIDAE	2	7
	FAMILY: ORUSSIDAE	3	13
	FAMILY: STEPHANIDAE	3	21
	FAMILY: MEGALYRIDAE	1	22
	FAMILY: TRIGONALIDAE	2	10
	FAMILY: CERAPHRONIDAE	3	51
	FAMILY: MEGASPILIDAE	4	24
	FAMILY: MYMAROMMATIDAE	2	3

WEBSITES

Atlas of Living Australia www.ala.org.au
Australian Faunal Directory www.biodiversity.org.au/afd
Australian Museum www.australianmuseum.net.au/insects
Brisbane Insects www.brisbaneinsects.com
Coffs Harbour Butterfly House www.lepidoptera.butterflyhouse.com.au
CSIRO www.csiro.au/en/Research/Collections/ANIC/ID-Resources
iNaturalist Australia www.inaturalist.ala.org.au
Insects of Tasmania www.insects-of-tasmania.org.au
Kape Images www.kapeimages.com.au
Museum and Art Gallery of the Northern Territory www.magnt.net.au
Museums Victoria www.museumsvictoria.com.au
Perth Museum www.museum.wa.gov.au
Peter Rowland Photographer and Writer www.prpw.com.au
Queensland Museum www.qm.qld.gov.au
South Australian Museum www.samuseum.sa.gov.au
Tasmanian Museum and Art Gallery www.tmag.tas.gov.au

REFERENCES

Braby, Michael F. (2016) *The Complete Field Guide to Butterflies of Australia* (2nd ed.) CSIRO Publishing, Vic.

Brock, P. D. & Hasenpusch, J. W. (2009) *The Complete Field Guide to Stick and Leaf Insects of Australia*. CSIRO Publishing, Vic.

Burwell, C., Monteith, G. B, Wright, S. & Peters, B. C. (2007) *Insects*. pp. 85–173. In Ryan, M., ed., *Wildlife of Greater Brisbane* (2nd ed.) Queensland Museum, South Brisbane, Qld.

Chaiwong, T., Srivoras, T., Sueabsamran, P., Sukontason, K., Sanford, M. R. & Sukontason, K. L. (2014) The blow fly, *Chrysomya megacephala*, and the house fly, *Musca domestica*, as mechanical vectors of pathogenic bacteria in Northeast Thailand. *Tropical Biomedicine* 31(2):336-46.

Department of Agriculture and Fisheries (2018) *A–Z List of Horticultural Insect Pests*. Queensland Government, on www.daf.qld.gov.au/business-priorities/plants/fruit-and-vegetables/a-z-list-of-horticultural-insect-pests.

Djernæs, M., Varadínová, Z., Kotyk, M., Eulitz, Ute & Klass, K.D. (2020) Phylogeny and life history evolution of Blaberoidea (Blattodea). *Arthropod Systematics and Phylogeny*. 78. 29-67.

Engel, Michael S. (2015) Insect evolution. *Current Biology* 25, R845–R875. Elsevier Ltd.

Farrow, R. (2016) *Insects of South-Eastern Australia: An Ecological and Behavioural Guide*. CSIRO Publishing, Clayton South, Vic.

Hangay, G. & Zborowski, P. (2010). *A Guide to the Beetles of Australia*. CSIRO Publishing, Vic.

Hausmann, A., Hebertm P. D. N., Mitchell, A., Rougerie, R., Sommerer, M., Edwards, T. & Young, C. J. (2009) Revision of the Australian Oenochroma vinaria Guenée, 1858 species-complex (Lepidoptera: Geometridae, Oenochrominae): DNA barcoding reveals cryptic diversity and assesses status of type specimen without dissection. *Zootaxa* 2239:1–21.

Houston, Terry. (2018) *A Guide to Native Bees of Australia*. CSIRO Publishing, Vic.

Manasria, T., Moussa, F., El-Hamza, S., Tine, S., Megri, R. & Chenchouni, H. (2014) Bacterial load of German cockroach (*Blattella germanica*) found in hospital environment. *Pathogens and Global Health* 108(3): 141–147.

Mantič M., Sikora T., Burdíková N., Blagoderov V., Kjærandsen J., Kurina O., Ševčík J. (2020) Hidden in Plain Sight: Comprehensive Molecular Phylogeny of Keroplatidae and Lygistorrhinidae (Diptera) Reveals Parallel Evolution and Leads to a Revised Family Classification. *Insects*. 2020 Jun 4;11(6):348

Matthews, R. W. & Matthews, J. R. (2009) Nesting behavior of *Abispa ephippium* (Fabricius) (Hymenoptera: Vespidae: Eumeninae): Extended Parental Care in an Australian Mason Wasp. *Psyche* Vol. 2009, Article 851694.